Scaling Decarbonisation Solutions: Reducing Emissions by 2030

Rotterdam, The Netherlands
29 November - 1 December 2022

ISBN: 978-1-7138-8302-9

Printed from e-media with permission by:

Curran Associates, Inc.
57 Morehouse Lane
Red Hook, NY 12571

Some format issues inherent in the e-media version may also appear in this print version.

The Royal Institution of Naval Architects
8-9 Northumberland Street
London, WC2N 5DA
United Kingdom

Phone: 020 7235 4622
Fax: 020 7259 5912

publications@rina.org.uk

Additional copies of this publication are available from:

Curran Associates, Inc.
57 Morehouse Lane
Red Hook, NY 12571 USA
Phone: 845-758-0400
Fax: 845-758-2633
Email: curran@proceedings.com
Web: www.proceedings.com

160 YEARS OF HISTORY, KNOWLEDGE AND EXPERIENCE

THE ROYAL INSTITUTION OF NAVAL ARCHITECTS

Scaling Decarbonisation Solutions:

Reducing Emissions by 2030

29th November – 1st December 2022, Rotterdam, The Netherlands

In Partnership with:

Mærsk Mc-Kinney Møller Center
for Zero Carbon Shipping

TOGETHER WE WILL SHAPE OUR FUTURE

THE ROYAL INSTITUTION
OF NAVAL ARCHITECTS

Scaling Decarbonisation Solutions:
Reducing Emissions by 2030

29th November – 1st December 2022
Rotterdam, The Netherlands

Papers published in these conference Proceedings are peer reviewed.

THE ROYAL INSTITUTION OF NAVAL ARCHITECTS
8-9 Northumberland Street
London WC2N 5DA

Telephone: 020 7235 4622
Fax: 020 7259 5912

ISSN No: 2752-6569

CONTENTS

BUILDING ON GREEN SHIPPING CORRIDORS: INTERNATIONAL BLOCS FOR ZERO-EMISSION SHIPPING

A Browne, Pacific Environment, USA

SUMMARY

This paper assesses the scalability of port-to-port green shipping corridors and suggests a collaborative model to build on corridor development. In addition to corridors, nations, sub-nationals and their ports can align with regional partners to create international blocs of high ambition ports. By leveraging economic buying power and public-private investments and aggregating demand for renewable electricity and zero-emission (ZE) fuels, regional alliances can develop infrastructure at scale, create clusters for innovation and promote rapid deployment of ZE ships. We examine arguments ports cite as concerns about decarbonisation policies, including cargo diversion, increased costs and reduced profits. By looking to existing examples of regional policy and multi-port collaboration, we challenge the assertion that first movers will suffer loss of business. Regional alliances can generate a rising tide – ensuring that ports of all sizes, economic niches and vessel specialisation can reduce emissions on a trajectory commensurate with our climate emergency.

NOMENCLATURE

CH_4	Methane
CO_2	Carbon dioxide
ETC	Energy Transitions Commissions for the Global Maritime Forum for the Getting to Zero Coalition
ETS	Emissions Trading System of the European Union
EU	European Union
Gt	Gigatons (or Gigatonnes)
ICCT	International Council on Clean Transportation
IMO	International Maritime Organization
IRA	U.S. Inflation Reduction Act
LNG	Liquefied natural gas
MARAD	U.S. Department of Transportation Maritime Administration
MoU	Memorandum of Understanding
N_2O	Nitrous oxide
NWPCAS	Northwest Ports Clean Air Strategy
NWSA	Northwest Seaport Alliance
PIDP	Port Infrastructure Development Program
UMAS	University Maritime Advisory Services
UNCTAD	United Nations Conference on Trade and Development
ZE	Zero-Emission

1. INTRODUCTION

Maritime shipping is a significant contributor to both climate-warming emissions and air pollution. If the industry were its own country, it would be the sixth largest polluter of greenhouse gas (GHG) emissions [1], including carbon dioxide (CO_2), methane (CH_4) and nitrous oxide (N_2O). Studies show that air pollution from shipping contributes to 250,000 premature deaths and 6.4 million cases of childhood asthma globally each year, with critical impact zones in China, Singapore, Panama and Brazil, and along the coastlines of Asia, Africa and South America [2]. Beyond nations' commitments under the Paris Agreement to keep global temperature rise below 1.5 to 2 degrees Celsius, nations have a moral imperative to accelerate the transition to zero-emission shipping to eliminate this lethal pollution for port communities and their regional air sheds. To account for maritime trade's exponential growth trajectory, the shipping industry must achieve zero GHG emissions by 2040 in order to align with a 1.5 degrees Celsius scenario [3]. The 2020s are a decisive decade for climate action, and nations, sub-nationals and their ports must step up to the challenge.

This paper assesses the scalability of port-to-port green shipping corridors to solve this urgent need for global action and suggests a collaborative model to build on corridor development. In addition to corridors, nations, sub-nationals and their ports can align with regional partners to create international blocs of high ambition ports. By joining together to take bold climate and public health actions, regional alliances can accelerate the zero-emission (ZE) shipping transition this decade. These international alliances of high ambition ports can reinforce the green shipping corridors developed under the COP26 Clydebank Declaration, helping to bring zero-emission shipping to ports around the world. We look to existing examples of regional policy and multi-port alliances as models for collaboration. Finally, we challenge the assertion that first movers will suffer cargo diversion, increased costs and reduced profits. Regional alliances can generate a rising tide – ensuring that ports of all sizes, economic niches and vessel specialisation can reduce emissions on a trajectory commensurate with our climate emergency.

Together with green shipping corridors, these international blocs of high ambition ports will send a clear market signal to the shipping value chain. The renewable electricity and ZE marine fuels market will scale to support aggregated demand across the region, ensuring reliable access to zero-emission electricity and fuel bunkering for ships. With sufficient ZE mandates in place, shipping companies will have no choice but to transition their fleet: retrofitting their current fleets with lifesaving and energy saving technologies and ordering zero-emission vessels at scale. Shipping's energy transition will drive billions of dollars of economic investment into ports and port communities, including infrastructure development and job creation, while simultaneously improving the health of local communities through reduced air, water and land pollution.

2. THE FOUNDATION: GREEN SHIPPING CORRIDORS FOR ZERO-EMISSION SHIPPING

2.1 NATIONAL COMMITMENTS TO GREEN SHIPPING CORRIDORS

Under the COP26 Clydebank Declaration, 24 countries acknowledged the "need for the formation of an international coalition between ambitious governments, to act together and demonstrate that maritime decarbonisation is possible, while unlocking new business opportunities and socioeconomic benefits for communities across the globe." The signatories committed to support the establishment of at least six green corridors by 2025, "while aiming to scale activity up in the following years". As of October 2022, signatories include Australia, Belgium, Canada, Chile, Costa Rica, Denmark, Fiji, Finland, France, Germany, Ireland, Italy, Japan, Republic of the Marshall Islands, Morocco, Netherlands, New Zealand, Norway, Palau, Singapore, Spain, Sweden, the United Kingdom of Great Britain and Northern Ireland and the United States of America. [4]

In June 2022, the United States and Norway announced their global Green Shipping Challenge for COP27. The initiative calls for stakeholders across the shipping value chain to "come forward with concrete steps that will help put the international shipping sector on a credible pathway this decade toward full decarbonisation no later than 2050." For ports, the initiative highlights the need for zero-emission bunkering and recharging capabilities to support low- or zero-emission vessels. The countries invited leaders to announce specific actions under the initiative at COP27 in November 2022 [5].

Globally, the message is clear: collaboration is key to success. With the foundation of an international coalition in place, nations and ports must do all they can to deliver on the promises made at COP26 and as required by the Paris Agreement.

2.2 GREEN SHIPPING CORRIDORS UNDER DEVELOPMENT

In less than a year, ports around the world announced green shipping corridor commitments encompassing some of the busiest global trade routes and key bunkering ports. COP27 will celebrate the one-year anniversary of the Clydebank Declaration and the leadership of these ports.

The Dutch Port of Antwerp and the Canadian Port of Montreal were the first ports to establish a green shipping corridor in the North Atlantic. The Port of Antwerp CEO Jacques Vandermeiren declared the port's intent to play a "pioneering role in the integration of carbon-neutral fuels in the bunkering market" – recognizing the need for key bunkering ports to lead the transition to cleaner fuels. Building on their framework partnership in place since 2013, the North American and European ports pledged to mobilize their respective public- and private-sector partners in the assessment, identification, development and adoption of shared or complementary solutions and infrastructures [6].

In January 2022, the Californian Port of Los Angeles and the Chinese Port of Shanghai were the first ports to announce a transpacific green shipping corridor [7]. Port of Los Angeles' sister Port of Long Beach joined the corridor in June 2022 [8], solidifying a green corridor commitment from the largest seaport complex in North America (Los Angeles-Long Beach). Together with key maritime stakeholders, the ports intend to release their "Green Shipping Corridor Implementation Plan" at COP27, including deliverables, milestones and roles for the partnership [9]. As one of the busiest trade routes in the world, the L.A.-L.B.-Shanghai corridor network will affect a significant number of vessels and countries that move goods between these ports.

A European corridor network was the next to emerge. With support from the Mærsk Mc-Kinney Møller Center for Zero Carbon Shipping, port authorities in Northern Europe and the Baltic Sea established the European Green Corridors Network in March 2022, including Hamburg, Germany; Gdynia, Poland; Rønne, Denmark; Rotterdam, the Netherlands and Tallinn, Estonia. The European Green Corridors Network aims to demonstrate the early commercialisation of alternative fuel supply chains, showcase and support first-mover solutions, and create a blueprint for rolling out green corridors in other areas and regions. Most importantly, it was the first regional network to materialise, though timing for its phased approach is unclear. The initiative will begin with a pre-feasibility phase to establish "high impact green corridors in the region" before moving into a technical, regulatory and commercial feasibility study of shortlisted routes and a final implementation phase [10].

Rotterdam added to its green corridor commitments in August 2022, signing a Memorandum of Understanding (MoU) with Singapore to establish the world's longest Green and Digital Corridor to enable low- and zero-carbon shipping. The coalition aims to bring together stakeholders from across the supply chain in an effort to raise investment confidence, attract green financing, and kick start joint bunkering pilots and trials for digitalisation [11].Vital to bunkering on Asian-European shipping routes, these ports' commitment to a green shipping corridor reinforces the importance of prioritizing infrastructure development to accelerate the zero-emission transition. If zero-emission ships have nowhere to bunker zero-emission fuels, the shipping value chain will be hard-pressed to purchase, build and deploy these ships. Ports have a responsibility to scale the infrastructure necessary to fuel the future of shipping.

2.3 ASSESSING GLOBAL SCALABILITY OF PORT-TO-PORT CORRIDORS

Port-to-port agreements are a step in the right direction for decarbonisation, but they may be difficult to scale globally. This is due in large part to how few shipping routes are direct, port-to-port routes (or direct service corridors). Only 2.9 per cent of the world's 12,748 port pairs have a direct service corridor. For trade between the remaining 97.1 per cent of port pairs, cargo must be transhipped in other ports [12]. In other words, green shipping corridor agreements for approximately 12,378 of the world's port pairs would need to involve more than two ports to ensure that all ports along that pair's corridor were aligned on zero-emission policy, infrastructure and bunkering. To date, we have four green corridor agreements for shipping, and only the European Green Corridor Network aligns more than three ports on decarbonising their shared corridors. To bring the rest of the world's ports and trade routes under green shipping corridors would be a complicated undertaking, to say the least.

With data from MDS Transmodal, the United Nations Conference on Trade and Development (UNCTAD) illustrated this port connectivity in two key figures, shown below. The figures describe each port's position in the shipping network by two metrics:

- The node *degree*: the number of other ports with which the port has a direct connection
- The node *betweenness*: How important is a port for trade between other ports that connect with each other via transhipment.

Figure 1 depicts the top 50 ports ranked by *degree* (all data is for the second quarters of each year). **Figure 2** shows the top 50 ports ranked by their *betweenness* [13].

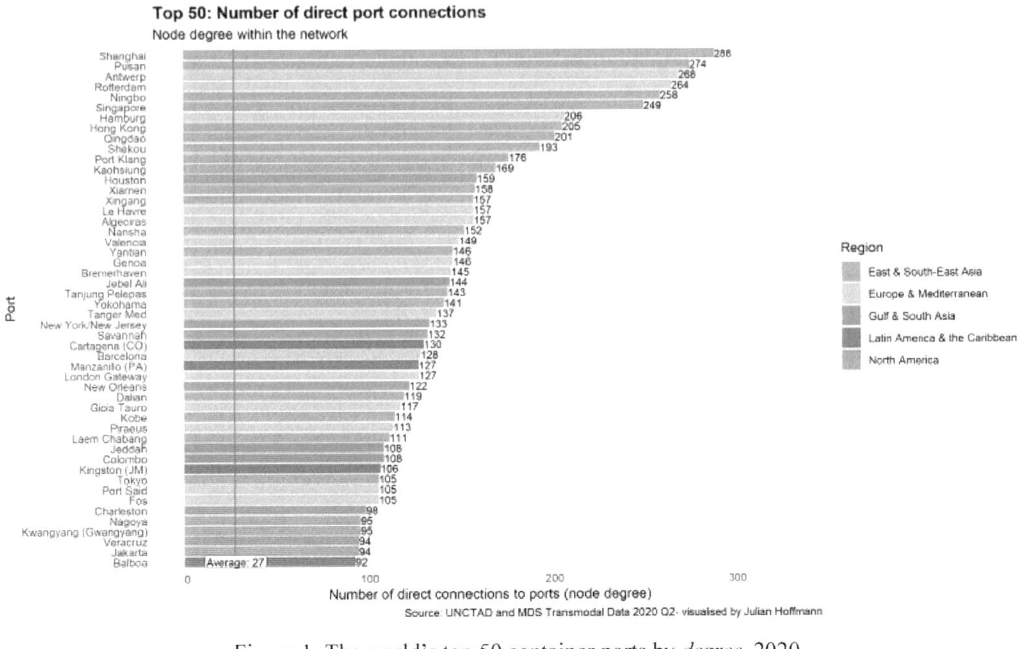

Figure 1: The world's top 50 container ports by *degree*, 2020

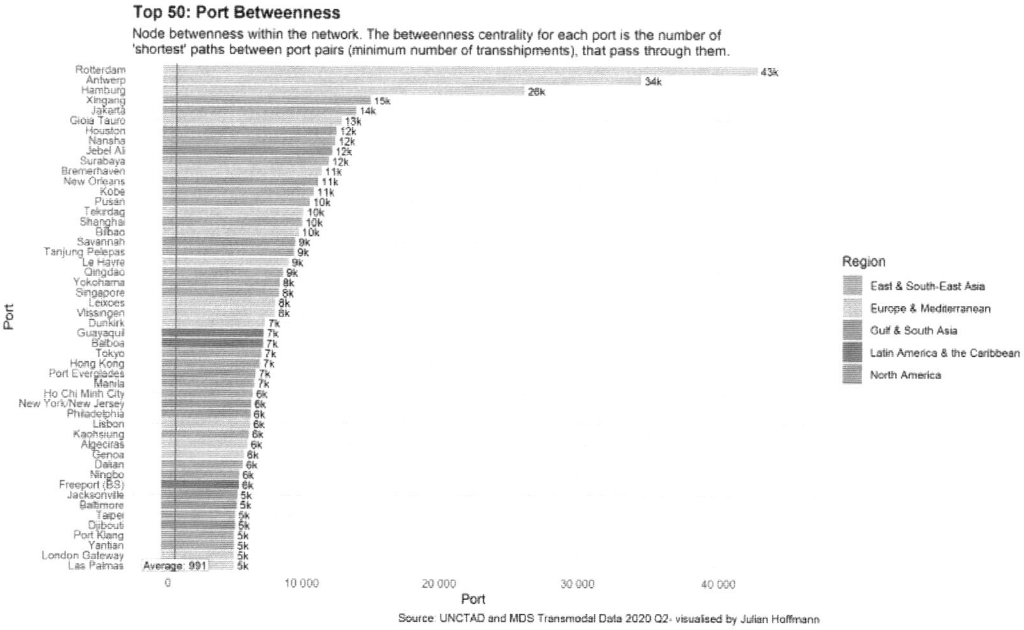

Figure 2: The world's top 50 container ports by *betweenness*, 2020

The study highlighted the Port of Rotterdam as the leader for *betweenness*, meaning it is a key stop for the highest number of indirect shipping routes. Approximately 43,000 shipping corridors include Rotterdam on their optimal routes, which is nearly 10 per cent of the global total of 440,391 port-to-port connections [14].

To put it simply, it would take thousands of multi-port green shipping corridor agreements to decarbonise the industry from a corridor approach. Given what we have seen in green shipping corridor development in the last twelve months, these corridors will take years to progress from concept to zero emissions. While ports should continue to pursue green shipping corridors as a means to jumpstart development of zero-emission infrastructure and fuels along the most heavily trafficked trade routes, ports of all sizes can simultaneously pursue regional collaboration to accelerate the global transition. This approach is our best chance of deep and swift industry decarbonisation.

2.4 DECARBONISATION IS A WHOLE SYSTEM CHALLENGE & OPPORTUNITY

Research by University Maritime Advisory Services (UMAS) and the Energy Transitions Commissions for the Global Maritime Forum for the Getting to Zero Coalition (ETC) shows that 87 per cent of the investments needed to decarbonise shipping are for land-based infrastructure and production facilities for low carbon fuels. On the study, Lord Adair Turner, Chair of the Energy Transitions Commission affirmed: "Much of shipping's decarbonisation will take place on land. It is a systemic transformation that goes beyond the capabilities of the maritime industry alone. We need to bring together the full range of upstream and downstream fuels value chains to unlock shipping's shift to zero carbon energy sources. Done right, this represents a trillion-dollar market opportunity" [15].

Ports can leverage public- and private-sector funding to stake a claim in this trillion-dollar market – positioning themselves as market makers and leaders in the clean energy transition for ocean shipping. Studies show that the U.S. [16], Morocco [17], Chile [18], South Africa [19], Mexico [20], Indonesia [21], and other countries around the globe have significant potential to develop renewable energy projects for zero-emission shipping fuel. For ports to engage on this development, they will need to collaborate with clean energy industry, utilities, regulators and the broader energy market. This is a tall order for ports of varying size and economic influence, but there are ways for ports to level up, and it starts with regional collaboration.

3. SCALING DECARBONISATION: INTERNATIONAL BLOCS FOR ZERO-EMISSION SHIPPING

3.1 REGIONAL POLICY AS A MODEL

3.1 (a) The European Union's ETS & FuelEU Maritime

In July 2022, the European Parliament adopted a historic proposal to include shipping emissions in the European Union (EU)'s Emissions Trading System (ETS), which is the bloc's cap-and-trade system [22]. The program's scope will include CO_2 emissions from large ships above 5,000 gross tonnage, regardless of ship flag. g. The extension will include "all emissions from ships calling at an EU port for voyages within the EU (intra-EU) as well as 50% of the emissions from voyages starting or ending outside of the EU (extra-EU voyages), and all emissions that occur when ships are at berth in EU ports" [23].

The European Union's (EU) Fit for 55 package includes the FuelEU Maritime proposal to reduce the greenhouse gas (GHG) intensity of energy of marine fuels. In October 2022, the European Parliament approved a target to reduce GHG intensity 80 per cent by 2050 from a 2020 baseline. The target applies to all energy used on ships involved in intra-EU traffic and 50 per cent of energy used for outbound and inbound international voyages [24].

Although the policies are pending finalisation through the EU trilogues – the inter-institutional negotiations for the adoption of EU legislation – we see how a bloc of nations committed to a common goal can target emissions from this sector via standardized policy and market mechanisms. Of course, the EU is unique in that it has created a robust structure for international governance and regulation, but regional alliances need not have this structure to adopt common goals and standardized policies for decarbonisation.

3.1 (b) U.S. Clean Shipping Act of 2022

In the United States, California Representative Alan Lowenthal introduced federal legislation that would mandate the shipping industry to achieve 100 per cent GHG reductions by 2040. Specifically, the bill directs the EPA to promulgate regulations by January 1, 2026, to reduce GHG emissions from marine vessels of 400 gross tonnage and above that call on ports in the United States. The bill also directs the EPA to require zero in-port emissions from marine vessels by 2030. Like the FuelEU Maritime policy, the U.S. Clean Shipping Act focuses on the carbon intensity of marine fuels, but the legislation requires decarbonisation on a faster timeline than EU – calling for 100 per cent reductions by 2040 [25].

If the Clean Shipping Act passes, ports could leverage existing funding designated by the U.S. Inflation Reduction Act (H.R. 5376, Sec. 60102). Included in the IRA is a USD 3 billion allocation for Grants to Reduce Air Pollution at Ports, plus an additional USD 1 billion to replace heavy-duty vehicles with zero-emission alternatives, including those that service ports [26]. Beyond the IRA, U.S. ports can already access funding through the U.S. Department of Transportation Maritime Administration (MARAD) Port Infrastructure Development Program (PIDP), which grants awards for environmental projects including electrification, hydrogen infrastructure and coastal resiliency [27].

This policy plus funding combination demonstrates how nations can support port infrastructure for zero-emission shipping while mandating a zero-emission transition from the industry at large. As regions consider how best to approach these alliances, nations must prioritize funding and financing to accelerate sustainable infrastructure development for renewable electricity, batteries and zero-emission marine fuel hubs.

3.2 REGIONAL PORT COLLABORATION IN THE PACIFIC NORTHWEST

Along the Northwest Pacific Coast of the U.S. and Canada, ports have joined forces to alleviate competitive pricing *and* to encourage environmental action above competition. This regional collaboration demonstrates the power and benefit of allying with domestic and international partners.

In 2014, the seaports of Seattle and Tacoma in Washington, U.S. "announced a bold vision: instead of competing, the two ports would join forces and form an alliance to better compete on a global scale." The ports established the Northwest Seaport Alliance (NWSA) in 2015, the first of its kind marine cargo operating partnership for North America. Under a port development authority, the NWSA manages the container, breakbulk, auto and some bulk terminals in Seattle and Tacoma. Today, NWSA is the fourth-largest container gateway in the U.S., handling USD 66 billion of waterborne trade with 189 trading partners globally in 2020. The Alliance collaborates with the Ports of Seattle and Tacoma – which still operate independently – to enhance the region's community by creating jobs, reducing the gateway's environmental footprint and boosting the state's economy [28].

Washington's key ports took their collaboration international with the Northwest Ports Clean Air Strategy (NWPCAS). First adopted in 2008, the strategy sought to "encourage environmental action above competition" to reduce seaport-related emissions that contribute to air pollution in the shared Puget Sound-Georgia Basin Air shed. The Northwest Seaport Alliance, Port of Seattle, Port of Tacoma and Port of Vancouver, British Columbia adopted the updated Northwest Ports Clean Air Strategy in 2021 with a vision to "phase out emissions from seaport-related activities by 2050, supporting clean air for our local communities and fulfilling our shared responsibility to help limit global temperature rise to 1.5°C." The initiative seeks to reduce emissions from ocean-going vessels, drayage trucks, cargo-handling equipment, rail, harbour vessels and port administration and facilities [29].

Partners of the NWPCAS initiative include the U.S. Environmental Protection Agency, the Washington State Department of Ecology, Puget Sound Clean Air Agency, Metro Vancouver, Province of British Columbia, Transport Canada, and Environment Canada, with input from non-governmental organizations, near-port community groups, industry, and local governments. The international initiative ensures that "environmental performances is not compromised but driven forward over and above competing interests" [30].

The Northwest Seaport Alliance and the NWPCAS demonstrate that allied action can enable ports to succeed both economically and environmentally to benefit the greater good. That is, ports need not choose between economic profitability and environmental responsibility. These alliances serve as a model for what green shipping corridors have the potential to become – an international network of port alliances committed to a zero-emission future for shipping.

3.3 ESTABLISHING REGIONAL ALLIANCES

International blocs of high ambition ports can complement and reinforce green shipping corridors. More importantly, addressing emissions from shipping on a regional level – much like the European Union is working to do with the European Trading Scheme and FuelEU Maritime – can provide stronger pollution and emissions reductions for regional air sheds and maximize the environmental and economic benefits of a zero-emission shipping industry.

3.3 (a) Economic and Environmental Benefits of a Regional Alliance on Zero-Emission Shipping

By leveraging economic buying power and public-private investments and aggregating demand for renewable electricity and zero-emission fuels, regional alliances can develop infrastructure at scale, create clusters for innovation and promote rapid deployment of zero-emission ships. Moreover, international collaboration will ensure that ports of all sizes, economic niches and vessel specialisation can reduce emissions on a trajectory commensurate with our collective climate emergency.

By advancing a regional commitment to achieve zero-emission shipping by 2040, nations, sub-nationals and their ports can:

- Establish common targets and policies to standardize regional compliance for ships.
- Maximize national funding earmarked for infrastructure development by demonstrating unified prioritisation to decarbonise shipping and reduce pollution.
- Accelerate adoption of electric shore power and zero-emission fuels.
- Send unified market signals to the cargo shipping value chain.
- Level the playing field for ports to progress in a collaborative environment.
- Identify market gaps for the region, creating economic opportunities for smaller ports, including handling specialty cargo or specialised vessels.
- Unite ambition across shipping and aviation for joint sea- and airport authorities for scalable decarbonisation across freight transportation sectors.

3.3 (b) Goals and Implementation

Once established, regional alliances must agree on a common mission and goals. The mission is already set accelerate ocean shipping decarbonisation and eliminate emissions by 2040 in order to align with a 1.5 degrees Celsius scenario. This target is based on shipping's proportional share of the remaining 1.5 degrees Celsius and 2 degrees Celsius carbon budgets of 10 Gigatons (Gt) and 17 Gt, respectively [31].

Each region should collaborate to achieve three key goals:

- Create an international, high ambition bloc of nations, sub-nationals and ports.
- Advance a common regulatory and funding framework to eliminate emissions from ocean-going vessels by 2040.

- Achieve measurable progress benchmarks for zero-emission ports and shipping, accounting for emissions from ships on route, on shore and in harbour.

Our *Ports Playbook for Zero-Emission Shipping* lays out three main tracks for ports to achieve zero-emission shipping: commitments, policy and progress. These tracks lay out nine actions to end port pollution, accelerate the market for zero-emission technologies, reward first movers and ensure reliable access to zero-emission infrastructure and fuels [32]. As nations, sub-nationals and ports work to develop their goals and implementation strategies for zero-emission shipping, we encourage a standardized set of commitments, policies and progress priorities like those in the *Ports Playbook*.

3.3 (c) Framework & Convention

Based on port ownership models and the necessitated involvement of regulatory and government agencies, there is no one-size-fits-all model for governing regional alliances. However, we recommend alliances to incorporate the following steps and processes into their framework, convening and implementation plans:

- Establish a forum for alliance communications.
- Hold an inaugural meeting for stakeholder alignment, goal setting and timeline planning.
- Integrate input and support from community, environmental justice, climate action and labour groups, and designate stakeholder representatives from each group.
- Set short-term (interim) targets to achieve measurable progress benchmarks on the path to zero emissions.
- Conduct ongoing initiative assessments and progress reports.

These steps and processes are critical to developing, implementing and achieving a just and equitable zero-emission transition for the region and its people.

3.3 (d) Priority Regions for Alliances

Any combination of countries within a region could ally to form an international bloc of high ambition ports. Taking into account the world's largest container ports, critical impact zones for air pollution caused by shipping, and regions ripe for economic and clean energy development, we have identified several regions critical to scaling decarbonisation, including:

- **The Western Pacific**, spanning from Alaska, U.S. through British Columbia, Canada, the U.S. states of Washington, Oregon and California to the Pacific Coast of Mexico.
- **The Eastern Pacific and Southeast Asia**, including key port nations of Japan, the Republic of Korea, Singapore, China, Malaysia, Thailand, Taiwan, Indonesia and Vietnam.
- **The North Atlantic and Baltic Sea**, including the ports of Western Europe and Scandinavia.
- **The Eastern Atlantic**, including Canadian and U.S. East Coast ports, including the Ports of Montreal, Canada; New York-New Jersey; and Savannah, Georgia.
- **Central and South America**, including key ports in Brazil, Panama, Colombia, Peru and Ecuador.
- **Southern Europe and the Mediterranean**, including key ports in Spain and Greece.
- **Africa and the Middle East**, including ports in Sub-Saharan Africa and the coasts of the Arabian Sea.

3.4 REGIONAL COLLABORATION ADDRESSES ECONOMIC CONCERNS ON DECARBONISATION

3.4 (a) Claims of Cargo Diversion Are a Red Herring

Ports around the world have hesitated and balked on climate action for myriad reasons, chiefly among them: fear of cargo diversion. Port authorities claim that adopting policies mandating zero-emission ships will divert cargo from their docks, resulting in reduced profits and economic harm to the port, its labourers and its community.

This argument fails to consider the opportunity costs of such evasive behaviour against the cost of complying with regulatory requirements. Ocean shipping is the most efficient and cost-effective method of international transportation for the movement of goods, moving 90 per cent of total cargo every year [33], meaning that transhipping cargo through indirect trade routes or alternative modes of transport (air, rail, or truck) would significantly increase the cost of getting those goods to their final destination. In the case of major direct and throughput trade routes, shipping companies would be unlikely to divert cargo to a neighbouring port unless the opportunity cost (increased cost of transhipment) was less than the increased cost of compliance at the primary port.

Analysis conducted by the European Commission as part of its Impact Assessment of the FuelEU Maritime legislative proposal support this principle. The assessment concluded that re-routing trips to neighbouring countries to evade FuelEU

Maritime requirements would be more costly than simply complying with the requirements [34]. In short, evading compliance by re-routing ships would be more costly than compliance and untenable as a long-term solution.

When considering standardised policies adopted on a regional scale, long-distance cargo diversion is even less probable. For example, rerouting Trans-pacific trade to Atlantic seaboards is highly unlikely to be economically feasible. Beyond the opportunity costs of transhipment, this argument ignores the significant logistical complexities of international re-routes, including new ocean and hinterland transport routes and potential international import complications.

3.4 (b) The Cost of Decarbonisation is High, but Negligible

Decarbonising the global ocean shipping industry is an expensive undertaking. UMAS and ETC estimate that full decarbonisation by 2050 would cost USD 1.4 trillion to 1.9 trillion [35].

Despite this hefty price tag, studies show that economies of scale in global supply chains can absorb the vast majority of increased transport costs to cargo companies. In 2022, the European Federation of Transport & Environment estimated what effect the EU's existing proposal to charge carbon pollution from ships, combined with the proposal to mandate small amounts of green e-fuel use by 2030, would have on container shipping pricing and consumer goods coming from China. The results debunked claims that industry decarbonisation would be prohibitively expensive and cause exorbitant price hikes for consumers – finding that running ships entirely on green hydrogen-based fuels (e-fuels) would add less than EUR 0.10 to the price of a pair of trainers, and up to EUR 8.00 for a refrigerator [36].

3.4 (c) Regional Collaboration Can Help End Fossil Fuel Investments This Decade

One of the most important benefits of regional alliances on shipping decarbonisation would be ending fossil fuel investments this decade, including investments in liquefied natural gas (LNG). LNG is not an emissions-reduction fuel; in fact, it is often worse on a well-to-wake basis than conventional fuels. LNG is 85-95 per cent methane, which is up to 86 times more potent than CO_2 on a shorter timescale [37]. Increasing the use of LNG as a marine fuel will exacerbate the industry's effect on global warming. Regional alliances can counter and stop this dangerous trajectory by ending LNG's future as a marine fuel, and shifting focus to renewable electricity, batteries, energy efficiency, wind propulsion and zero-emission marine fuels, including green hydrogen-based fuels like green methanol and liquid green hydrogen for fuel cells.

Luckily, existing fleets can be retrofitted or accommodate green hydrogen fuels, in addition to new dual-fuel ships that will replace the most heavily polluting ships on the water. Studies from the International Council on Clean Transportation found that 99 per cent of the voyages made along the U.S.-China container shipping container corridor could be powered by green hydrogen, with only minor changes to fuel capacity or operations. For example, ships would need retrofits to replace 5 per cent of cargo space with hydrogen fuel storage or to add one additional port of call to refuel [38].

4. CONCLUSIONS

Building on the success of green shipping corridors, international blocs of high ambition ports formed under regional alliances have the potential to accelerate the transition to zero-emission shipping this decade. Given the urgency of our climate, ecological and public health crises, the trajectory of global shipping growth, and the importance of a sustainable shipping economy, zero-emission shipping is a moral imperative for all nations, sub-nationals and ports.

5. ACKNOWLEDGEMENTS

Pacific Environment is a global environmental organisation with a mission to protect communities and wildlife of the Pacific Rim. We support community leaders to fight climate change, protect the oceans, build just societies and move from fossil fuels toward a green economy. Pacific Environment has a consultative status at the International Maritime Organization (IMO).

In September 2022, Pacific Environment launched its **Ports for People** campaign to end port and ship pollution. Together with local communities, allies and partners, Ports for People seeks to transform ports from hotspots of fossil fuel pollution to thriving hubs of sustainable economic development and environmental protection – where clean tech innovations enable port communities to benefit from renewable energy, clean air and clean oceans, and provide a safe home for people and wildlife to live in harmony. Learn more at www.portsforpeople.org.

This report builds on our ***Ports Playbook for Zero-Emission Shipping***, which calls on ports to take action across three lines of efforts: commitments, policy and progress. These tracks lay out nine actions that will end port pollution, accelerate the market for zero-emissions technologies, reward first movers and ensure reliable access to zero-emissions infrastructure and fuels.

The author appreciates the support of her team at Pacific Environment, including strategic guidance, policy, research and editorial support from Madeline Rose, Climate Campaign Director; Teresa Bui, State Climate Policy Director; and Antonio Santos, Federal Climate Policy Director. We are grateful to the many port communities, port authorities and shipping and maritime experts who provide valuable insights to inform our *Ports for People* campaign. We would also like to acknowledge the impactful work of our campaign partners and allies, including Opportunity Green, Friends of the Earth-U.S., World Wildlife Fund-Mexico, the International Council on Clean Transportation (ICCT), the European Federation for Transport & Environment, Ocean Conservancy, Interamerican Association for Environmental Defense, member organisations of T.H.E. Impact Project and the Moving Forward Network.

6. REFERENCES

1. SCHLANGER, Z., 'If shipping were a country, it would be the world's sixth-biggest greenhouse gas emitter', *Quartz*, 18 April 2018.

2. UNIVERSITY OF DELAWARE, 'New fuel standards will decrease childhood asthma cases: Most significant improvement in global fuel standards for the shipping industry in 100 years, will also reduce number of premature deaths', *Science Daily*, 07 February 2018.

3. COMER, B., 'Zero-emission shipping and the Paris Agreement: Why the IMO needs to pick a zero date and set interim targets in its revised GHG strategy', *ICCT*, 08 September 2021.

4. 'Clydebank Declaration for Green Shipping Corridors', *COP26*, 2021.

5. 'Fact Sheet: President Biden to Galvanize Global Action to Strengthen Energy-Security and Tackle the Climate Crisis through the Major Economies Forum on Energy and Climate', *The White House Briefing Room*, 17 June 2022.

6. 'The Port of Antwerp and the Port of Montreal pledge to create a green shipping corridor', *American Journal of Transportation*, 04 November 2021.

7. 'Port of Los Angeles, Port of Shanghai, and C40 Cities announce partnership to create world's first transpacific green shipping corridor between ports in the United States and China', *C40 Cities*, 28 January 2022.

8. 'Port of Long Beach joins the Green Shipping Corridor', *Port of Long Beach*, 09 June 2022.

9. 'Port of Los Angeles, Port of Shanghai, and C40 Cities announce partnership to create world's first transpacific green shipping corridor between ports in the United States and China', *C40 Cities*, 28 January 2022.

10. CHAMBERS, S., 'Mærsk Mc-Kinney Møller Center for Zero Carbon Shipping launches port partnership to establish the European Green Corridors Network', *Splash247.com*, 30 March 2022.

11. 'World's Longest Green Corridor Planned between Singapore and Rotterdam', *The Maritime Executive*, 02 August 2022.

12. HOFFMAN, JAN; HOFFMAN, JULIAN, 'Ports in the global liner shipping network: Understanding their position, connectivity, and changes over time', *UNCTAD Transport and Trade Facilitation Newsletter N°87, Third Quarter 2020, Article No. 57*, 10 August 2020.

13. *Id.*

14. *Id.* Footnote [i] The network is considered symmetric, so the total number of bilateral connections is 939 x (939-1) / 2 = 440,391.

15. UMAS; ETC, 'New Study by UMAS Shows that Decarbonisation of the Shipping Sector is a Whole System Challenge and Not Something Just for Shipping', *UMAS*, January 2020.

16. GEORGEFF, E., ET AL., 'Scaling U.S. Zero-Emission Shipping: Potential Hydrogen Demand at Aleutian Islands Ports', *ICCT*, 22 June 2022.

17. ASH, N.; SCARBROUGH, T., 'Sailing on solar: Could green ammonia decarbonise international shipping?', *Environmental Defense Fund*, May 2019.

18. ASH, N., ET AL., 'Electrofuels for shipping: How synthetic fuels from renewable electricity could unlock sustainable investment in countries like Chile', *Environmental Defense Fund*, 2019.

19. ENVIRONMENTAL DEFENSE FUND; RICARDO ENERGY & ENVIRONMENT, ET AL., 'South Africa: fueling the future of shipping', *Global Maritime Forum, P4G Getting to Zero Coalition Partnership*, June 2021 [Revised June 2022].

20. CARPENTER-LOMAX, O., MARTINEZ, V.; WILKINSON, G.; ENVIRONMENTAL DEFENSE FUND; RICARDO ENERGY & ENVIRONMENT, 'Mexico: fuelling the future of shipping', *Global Maritime Forum, P4G Getting to Zero Coalition Partnership*, September 2021 [Revised June 2022].

21. CARPENTER-LOMAX, O., MARTINEZ, V.; WILKINSON, G.; ENVIRONMENTAL DEFENSE FUND; RICARDO ENERGY & ENVIRONMENT, 'Indonesia: fuelling the future of shipping', *Global Maritime Forum, P4G Getting to Zero Coalition Partnership*, November 2021 [Revised June 2022].

22. SCHULER, M., 'EU Parliament votes to include shipping in EU carbon market', *gCaptain*, 22 June 2022.

23. 'Reducing emissions from the shipping sector', *The European Commission*.

24. BOCKMANN, M.W., 'EU Parliament passes world's first marine fuel targets for decarbonisation', *Lloyd's List*, 19 October 2022.

25. H.R. 8336 – Clean Shipping Act of 2022, *U.S. Congress*, 12 July 2022 [introduced].

26. MANDRA, J.O., 'US ports eye $3bln for electrification from Inflation Reduction Act', *Offshore Energy*, 09 August 2022.

27. 'Port Infrastructure Development Grant Information', *U.S. Department of Transportation, Maritime Administration (MARAD)*.

28. NORTHWEST SEAPORT ALLIANCE, 'About Us', *nwseaportalliance.com*.

29. NORTHWEST SEAPORT ALLIANCE, 'Northwest Ports Clean Air Strategy', *nwseaportalliance.com*.

30. *Id.*

31. COMER, B., 'Zero-emission shipping and the Paris Agreement: Why the IMO needs to pick a zero date and set interim targets in its revised GHG strategy', *ICCT*, 8 September 2021.

32. BROWNE, A.; O'LEARY, A., 'Ports Playbook for Zero-Emission Shipping', *Pacific Environment*, https://portsforpeople.pacificenvironment.org/policymaker/ports-playbook/, 20 September 2022.

33. 'Ocean shipping and shipbuilding', *OECD*, https://www.oecd.org/ocean/topics/ocean-shipping/.

34. 'Impact Assessment Accompanying the Proposal for a Regulation of the European Parliament and of the Council on the use of renewable and low-carbon fuels in maritime transport: Annex 4: Analytical methods, 7. Methodological approach and results of the risk of carbon leakage analysis', *European Commission*, 14 July 2021.

35. UMAS; ETC, 'New Study by UMAS Shows that Decarbonisation of the Shipping Sector is a Whole System Challenge and Not Something Just for Shipping', *UMAS*, January 2020.

36. 'Green shipping would add just 8 cents to a pair of Nikes', *Transport & Environment*, 28 June 2022.

37. PAVLENKO, N., ET AL., 'The Climate implications of using LNG as a marine fuel', *ICCT*, 28 January 2020.

38. MAO, X., ET AL., 'Refueling Assessment of a Zero-Emission Container Corridor between China and The United States: Could Hydrogen Replace Fossil Fuels?', *ICCT*, 03 March 2022.

7. AUTHOR'S BIOGRAPHY

Allyson Browne serves as Pacific Environment's Climate Campaign Manager, Ports, leading their *Ports for People* campaign to end port and ship pollution and emissions. Allyson has more than ten years of experience in renewable energy, climate, and sustainability, including expertise in sustainable policy development and implementation, campaign and program management, and impactful research and data analysis. Allyson previously served as Director of Regulatory Affairs & General Counsel for SRECTrade, where she collaborated with solar industry and grassroots organizations to advocate for renewable energy policies in U.S. states. As Chief of Staff to Jared Blumenfeld for California Governor Gavin Newsom's 2018 gubernatorial campaign, Allyson led a team to develop policies on a range of climate and energy issues. She has contributed to climate-related projects in fields ranging from renewable energy and energy efficiency in the cannabis industry to developing a global database of climate movements with Climate 2025.

Allyson holds a Juris Doctor degree from Golden Gate University School of Law, with a focus on Environmental Law and on EU & French Law studies at the Université Paris Nanterre in Paris, France. She earned her bachelor's degree in Business Administration from the University of Southern California, Marshall School of Business, with a concentration in International Business and a Spanish Language Minor from the College of International Studies in Madrid, Spain.

HULL AIR LUBRICATION: A PRACTICAL, PROVEN ENERGY SAVING SOLUTION

A Dubois, Silverstream Technologies, UK

SUMMARY

Over the past decade, hull air lubrication has proven itself as a practical and scalable decarbonisation solution for the shipping industry. Air lubrication aims to reduce the frictional resistance of a ship, thereby cutting fuel burn and emissions. The unique system developed by Silverstream Technologies generates a microbubble carpet through fluid shearing and, since the company's technology came to market, has demonstrated an average of 5-10% net savings as such showing that effective energy saving technologies can be commercialised and scaled rapidly. The company's value proposition spans seven key areas: solution, performance, growth, people, capability, lifetime partner, and research and development, which will be explored in this paper.

1. INTRODUCTION

The shipping industry is faced with huge challenges to keep up with the ever-increasing global demand for transport. In 2021 alone, over 1,200 newbuild ships were ordered, adding to an internationally trading global fleet of over 60,000 vessels. Simultaneously, governments, regulatory bodies, and society are increasing their calls for substantial reductions of the industry's environmental footprint. Truly scalable solutions to reduce, and eventually eliminate, shipping-related carbon emissions will have to encompass both the newbuild market and existing fleet. Furthermore, such technologies should be flexible to suit the many different ship types, sizes, and operational profiles, while also having the ability to rapidly deploy standardised solutions.

Silverstream's solution is a hull air lubrication system (ALS) that harnesses fluid shearing to create a microbubble carpet beneath a vessel. The solution has shown to be a practical, proven, and scalable energy saving solution that can play its part among a myriad of technologies in the maritime industry's efforts to reduce shipping's environmental footprint. Silverstream has become the undisputed market leader in air lubrication technology and is helping the global shipping industry on a daily basis to comply with ever more stringent efficiency regulations and to decarbonise at pace. The company's value proposition can be formulated in seven key areas: solution, performance, growth, people, capability, lifetime partner, and research and development (as illustrated in Figure 1).

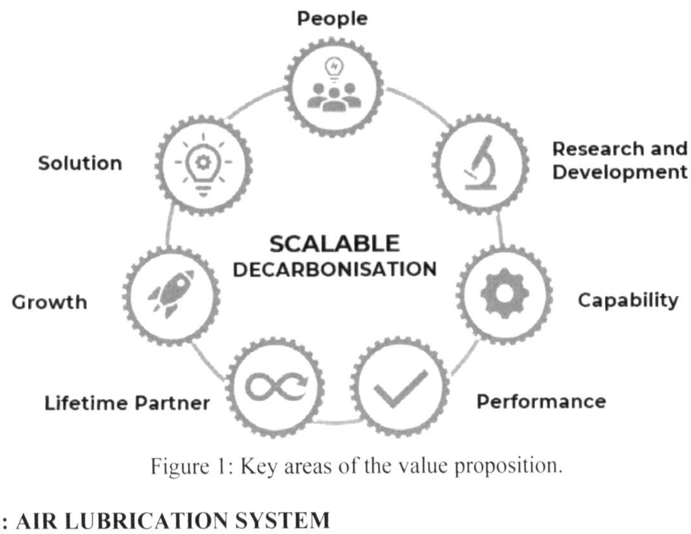

Figure 1: Key areas of the value proposition.

2. SOLUTION: AIR LUBRICATION SYSTEM

Silverstream's air lubrication system deploys a uniform carpet of microbubbles along the flat bottom area of a vessel, as such reducing the frictional resistance of the ship, allowing reduced fuel consumption and related emissions [1, 2, 3]. The system consists of four components: compressors, pipes and valves, Air Release Units (ARUs), and a control and automation system. Air is delivered through oil-free positive displacement screw compressors that are sized based on the specific vessel's operational profile and requirements (including specification of the number of compressors). The piping and valve system is further designed for each specific vessel. Air is introduced into the water from the patented ARUs through fluid shearing, a unique feature of the system, which requires less energy than other expulsion-based systems and produces high 'quality' microbubbles without disturbing the boundary layer on the ship hull. The open water interface inside the ARU allows for generation of microbubbles through the Kelvin-Helmholtz instability effect. The stability and effectiveness of the shearing mechanism will depend on the ship speed and draught and is controlled through the airflow provided. A row of ARUs is arranged along the flat bottom line at the front of the ship, maximising the affected surface area to achieve maximum frictional resistance reduction. The solution was developed through extensive experimental testing over a 10-year period at HSVA's HYKAT cavitation tunnel allowing exploration, evaluation, and validation of the design at full scale (avoiding the need for Reynolds scaling, essential when considering frictional resistance reduction applications like air lubrication). Test campaigns spanned a speed range from 6 to 18 knots and have included investigation of the interaction of ALS with different anti-fouling solutions. Based on the experimental testing, the control and automation system controls and optimises the airflow delivered as governed by the ship's operational conditions, i.e. ship speed and draught.

An important aspect of air lubrication systems as an energy saving solution is the fuel agnostic character. Because alternative or transformative fuels are less energy dense (thus requiring larger fuel volumes), and generally more expensive, the system can achieve energy savings independent of the fuel type used. Addressing energy consumption (and related emissions) on a fundamental level by tackling the ship's resistance provides a clear competitive advantage and is agnostic to all decarbonisation pathways. Additionally, owners are offered a flexible solution as the installation is simple and straightforward for both newbuild vessels and retrofits.

3. PERFORMANCE: IMPORTANCE OF DEMONSTRATED SAVINGS

Any new technology introduced into the market making claims of energy savings will have to demonstrate those savings as part of the development process, as well as to ensure adoption. The requirement to demonstrate savings in a clear and transparent manner will come from contractual obligations and is often needed to comply with regulations. Most importantly, though, this is needed to instil and strengthen market confidence in the technology. Such demonstration, which publication happens always in agreement with the ship owner, should also include detailing of the used test procedure and data analysis protocol to ensure reproducibility and accountability of the achieved results.

The performance of the ALS can easily be demonstrated by comparing the vessel performance with and without the system switched on [3, 4, 5]. Once the system is active the vessel's shaft power reduces, and the ship speed increases. The total gross savings can then be calculated by summing the shaft power reduction and the equivalent power as a result from the speed increase, while net savings are calculated by subtracting the system's power consumption from the gross savings. As such, the ALS will positively affect the ship's speed power curve, as illustrated in Figure 2.

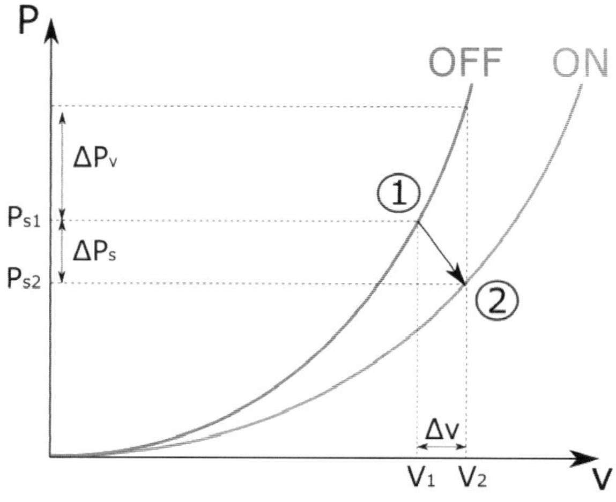

$$\Delta P_{net} = \Delta P_{gross} - \Delta P_{comp} \text{ with } \Delta P_{gross} = \Delta P_s + \Delta P_v$$

Figure 2: Savings achieved by ALS based on speed-power curves (gross savings ΔPgross, shaft power reduction ΔPs, equivalent power for increased ship speed ΔPv, net savings ΔPnet, and power consumed by ALS ΔPcomp).

Performance demonstrations have been conducted in various conditions, ranging from sea trials for first-of-class vessels, while other options exist for in-service performance monitoring such as straight-line trials as well as the system's internal automated self-test mode. The long-term performance analysis approach seeks to derive the speed-power relationship of the vessel for both the system switched on and off, allowing to demonstrate the impact of ALS on the ship's operation, enabling verification of the impact of external circumstance such as weather. The average net savings achieved by the ALS range between 5-10% and have independently been verified by various third parties and organisations such as RINA, Lloyd's Register, and HSVA [6, 7 , 8, 9], as well as by customers and academic institutions, confirming reported savings on a number of vessels / vessel types.

4. GROWTH: AIR LUBRICATION PIONEER AND MARKET LEADER

For over 10 years, Silverstream has pioneered the commercial air lubrication space [4]. The company has grown supported by ever-increasing regulatory pressure, spearheaded by the International Maritime Organization (IMO) ambitions in the fight against climate change in the form of EEXI, EEDI, and CII. A 'first movers' strategy was essential in the initial days of the company, with partners like Shell, Carnival, Grimaldi, Vale and MSC as thought leaders in each industry segment (e.g. LNGCs, cruise, roll-on-roll-off, bulker, container, etc.), establishing trust in the solution and ordering systems.

The company's continued growth was later accelerated by others following a 'close second' strategy and fleet deals and repeat orders from existing customers. In recent years, it has been demonstrated that air lubrication technology can be offered on a true commercial scale. Key to that success has been its clear and honest communication strategy, estimating realisable and realistic savings and demonstrating those savings in service. Furthermore, the aim is to be an energy saving thought leader has allowed the company to engage in strategic relationships with other industry thought leaders on various levels. The strategic relationships, spanning shipping companies, owners / operators / charterers, shipyards, and recently drydocks, allow to solidify its reputation, and build momentum to become an air lubrication market leader.

5. PEOPLE: GROWING THE TEAM

A diverse collection of professionals hailing from over 20 different countries, spanning marine engineering, data science, operations, and many other disciplines, collaborate together across Silverstream's offices in London and Shanghai to service the rapidly rising demand for its system. The company is building its Executive team to provide strategic oversight to its expansion plans. The latest addition to the team will be a Chief Data Officer in early 2023 who will lead the existing team of data scientists, data engineers and software engineers to harness the true power of the data accrued by its system while in-service onboard ships. Through extended analysis of the data generated, the system's performance can continuously be optimised and further research into improvements can be motivated by providing insights into various aspects of the system's operation. Across the Data team and the wider organisation, the company blends highly experienced professionals with graduates and early-stage professionals to help foster a culture of innovation and collaboration, which underpins everything the business does.

6. CAPABILITY: DELIVERING 500 SYSTEMS BY 2025

Through an excellent delivery record, a world class international supply chain with both depth and breadth through supply partners of significant reach and credibility, and global coverage, the organisation established itself as a trusted partner. From the start, it has focused on delivering performance-driven cost-efficient installations for owners and operators across all major shipping segments with at the present day a fleet of 24 vessels in service (projected at 28 vessels by end of 2022) and an orderbook for over 110 systems (across vessel types).

The company has set out the objective to deliver 500 systems by 2025 and keep growing beyond that, exceeding the number of systems delivered multifold. In order to realise these ambitions, a different approach is needed to expand the company's capabilities and play a key role in maritime decarbonisation efforts. As such, Silverstream is working on standardisation of the solution and on the introduction of application engineering principles. Through these processes, the decarbonisation solution is offered as standard products minimising design and project management efforts, while application engineers work directly with customers to understand specific needs and technical requirements, maintaining the company's reputation as the partner of choice in the space. Combined, these processes allow the further scale-up of the decarbonisation solution and achieve the growth objectives expressed.

7. LIFETIME PARTNER: DEDICATED AFTERSALES SUPPORT

The journey of the ship owner / operator with the organisation does not end at the system commission and installation, as the low maintenance Silverstream's ALS is intended to last the lifetime of the vessel, and the engagement between both parties continues throughout. Through extended interaction, aftersales and warranty services, training, system and ship performance evaluation and analysis, as well as system upgrades, the company aims to include its solution in the ship daily operation from the start and aims to ensure customers get as much use and value out of their installation as possible.

Additionally, the lifetime partnership encourages brand loyalty and opens up the potential for repeat orders and fleet deals, further scaling up the decarbonisation solution.

The company's commitment to be a lifetime partnership consists of three pillars: dedicated support service, in-service data analysis, optimisation, and continued system development. First, by aiming to offer dedicated support, it ensures increased system availability and reduce system downtime, supported by a world-leading supply and maintenance network. Second, in-service data analysis aims to report on the system performance and provide visibility and transparency of the savings achieved, while also offering insights into the general vessel performance. Finally, continued optimisation and system development, with a focus on the existing customer base and through extensive research and development programmes, aims to provide customers with an increasingly efficient and effective decarbonisation solution.

8. RESEARCH AND DEVELOPMENT: INNOVATOR AND THOUGHT LEADERSHIP

The company strives to continually innovate in the air lubrication technology space and demonstrate thought leadership overall within the green tech space, e.g. by chairing the Global Industry Alliance to Support Low Carbon Shipping. Silverstream is deepening its understanding of air lubrication, from both a physics and application perspective, pushing the boundaries of the technology and understanding on various fronts.

Currently, the continued development of the company's ALS is focused on the existing customer base, aiming to optimise performance in a cost-efficient way through potential retrofits and upgrades. A key focus in this process is the application of Machine Learning to maximise system performance in context of the ship operational orientation and encountered met-ocean conditions. The development of Machine Learning, whilst an in-house initiative, is being executed in collaboration with Southampton University and a number of the ship owners [5]. Additional research areas that are being pursued encompass, continued experimental testing (at HSVA), and the development of experimental approaches to study the interaction with coating and anti-fouling products in collaboration with industrial partners and the University of Newcastle. Furthermore, the company is conducting experimental and numerical research into the physics behind the bubble carpet generation, formation, and interaction, in addition to full-scale experimental and in-service measurements of the bubble carpet characteristics using sensing technology and the development of a Computational Fluid Dynamics model to allow for the evaluation of relative design changes and support further performance analysis. The company is also engaging in research projects with customers and industry partners, e.g. the CHEK project (EU Horizon 2020) which works on future-proof vessel platforms to radically reduce maritime energy consumption and emissions. Finally, work is conducted to explore and develop alternative airflow delivery systems in the form of scavenged air (SAALS) together with Maersk and Wartsila, as well as optimised compressor-ARU lay-outs to improve the system efficiency while reducing the cost the system.

9. CONCLUSIONS

The Silverstream® System is a scalable decarbonisation solution that has proven to offer energy savings for the shipping industry in an efficient and effective manner. The provider has established itself as the partner of choice for ship owners and operators in the application of air lubrication as a proven energy saving technology. Spanning seven key areas, it has set out a clear vision and value proposition. Over the last decade, Silverstream has become the market-leader in the air lubrication space with exponential 'growth' realised by a dedicated team of 'people', consisting of a mix of experienced professionals and new graduates. The company's ambitions to deliver 500 systems by 2025 are supported by 'capability' building through a global and robust supply chain, extensive standardisation to continue to deliver on set-out objectives, dedicated aftersales as 'lifetime partner' ensuring the relationship between them and the customer does not end at the system commission and installation, all underpinned by an extensive 'innovation, research and development' programme aimed at providing an increasingly efficient and effective decarbonisation solution.

10. REFERENCES

1. SILBERSCHMIDT, N., 'Air Lubrication: Setting the Industry Benchmark in Driving Sustainability and Operational Efficiency Through Proven Technology', *Energy Efficient Ship – RINA*, 23-24 November 2016

2. SILBERSCHMIDT, N., TASKER, D., PAPPAS, T., JOHANNESSON, J., 'Silverstream® System –Air Lubrication Performance Verification and Design Development', 10th Symposium on High-Performance Marine Vehicles (HIPER), 17-19 October 2016

3. SILBERSCHMIDT, N., PAPPAS, T., CONNOLLY, D., DE FREITAS, L., 'Full Scale Performance Measurement and Analysis of the Silverstream Air Lubrication System', *Full Scale Ship Performance – RINA*, 24-25 October 2018

4. CONNOLLY, D., DUBOIS, A., 'Status of Air Lubrication Technology', 14th Symposium on High-Performance Marine Vehicles (HIPER), 29-31 August 2022

5. CAMILLERI, J., SOBEY, A., CONNOLLY, D., HUDSON, D., 'Quantifying the Effect of Green Technologies on Ship Performance using Operational Data –Design of Experiments and Data Collection', 14th Symposium on High-Performance Marine Vehicles (HIPER), 29-31 August 2022

6. HSVA, 'Report ST108/17 Performance in Service Analysis', Technical Report, September 2017

7. LLOYD' REGISTER – SHIP PERFORMANCE GROUP, 'Independent Data Analysis of Air Lubrication System Performance', Technical Report, May 2018

8. RINA SERVICE S.P.A.., 'M/v Eco Valencia Air Lubrication System Performance Assessment', Technical Report, August 2018

9. HSVA, 'Verification of ALS Performance Test Calculations', Technical Report, May 2022

11. AUTHORS BIOGRAPHY

Arno Dubois holds the current position of IR&D Engineer (Hydrodynamics) at Silverstream Technologies. He is responsible for oversight and delivery of research-based projects aimed to analyse, understand, and improve the Silverstream® System. He deploys research techniques and practices in the fields of fundamental experimental research, computational fluid dynamics, underwater radiated noise, cavitations, execution of sea trials, and various performance measurement methodologies. His previous experience includes working as an Application Engineer for a simulation software company, specialising in computational fluid dynamics and heat transfer, and numerical and experimental research of hydrodynamic systems as part of his education and PhD research, which has been presented at several international conferences and published in various industry-leading journals.

QUANTITATIVE RISK ASSESSMENT OF AMMONIA FUELLED VESSELS

A P Franks, Lloyd's Register Maritime Decarbonisation Hub, UK
C Graugaard, Maersk McKinney Moller Center for Zero Carbon Shipping, Denmark

SUMMARY

A number of 'green' fuels are under active consideration by the maritime industry, including ammonia. Although shipping has extensive experience of designing, building and operating vessels that operate on hydrocarbon fuels, ammonia's toxicity presents a new hazard and raises serious safety concerns. Quantitative risk assessment (QRA) is an analytical tool that has been widely used in other industries, but limited use in the maritime sector to date. The technique can provide valuable insights into the risk profile of the systems studied, leading to recommendations for safety improvements. In a recent collaborative study, QRA has been applied to three ammonia-fuelled vessel reference designs: a containership, tanker and bulk carrier with different storage systems – fully refrigerated, semi pressurized and fully pressurized. This paper provides an outline of the QRA method used and presents examples of the results obtained.

Phase 2 of the study has now begun, as outlined in the paper.

NOMENCLATURE

$F_{s,o}$	Frequency of scenario s with outcome o [y^{-1}]
$IRPA_{C,l}$	Individual risk per annum for member of crew group C at location l [y^{-1}]
$IRPA_{C,tot}$	Total individual risk per annum for member of crew group C [y^{-1}]
$L_{F,C,l}$	Fraction of time spent at location l by member of crew group C while onboard [-]
$LSIR_{l,s,o}$	Location specific individual risk at location l for scenario s with outcome o [y^{-1}]
$LSIR_{tot,MUS}$	Total location specific individual risk at muster location l [y^{-1}]
$LSIR_{tot,l}$	Total location specific individual risk at location l [y^{-1}]
$MOD_{F,TP}$	Fraction of time vessel is in operating mode MOD during time period TP [-]
N_C	Number of personnel in crew group C [-]
$N_{l,TP,MOD}$	Number of personnel at location l during time period TP and operating mode MOD [-]
$P_{aff,l,s,o}$	Probability the location l is affected by scenario s and outcome o [-]
$P_{fail,esc}$	Probability that crew group member fails to escape from location [-]
$P_{fat,s,o}$	Probability of fatality from scenario s and outcome o [-]
PLL_{tot}	Total PLL for vessel as calculated using method 1 [fatalities.y^{-1}]
PLL'_{tot}	Total PLL for vessel as calculated using method 2 [fatalities.y^{-1}]
$PLL_{TP,l}$	PLL contribution during time period TP at location l [fatalities.y^{-1}]
TP_F	Fraction of time for time period TP [-]
$Y_{F,C}$	Fraction of year spent onboard ship by members of crew group C [-]

1. INTRODUCTION

1.1 THE PROJECT

The work described in this paper was performed as Phase 1 of the Ammonia Safety Project, a collaborative initiative involving the Lloyd's Register Maritime Decarbonisation Hub (LRMDH), the Maersk McKinney Møller Center for Zero Carbon Shipping (MMMCZCS), Maersk, Mitsubishi Heavy Industries, NYK Line, TotalEnergies and MAN Energy Solutions.

The objectives of the Phase 1 project were to:

- Provide an estimate of the risk of fatality to crew members working onboard ammonia-fuelled cargo ships;
- Determine the acceptability of these risks in relation to established criteria; and,
- Propose measures to reduce the risks where needed.

1.2 AMMONIA AS A POTENTIAL ZERO-CARBON FUEL

The global energy transition drives a move from fossil fuels to alternative energy sources. This brings new safety challenges and a need for shipping to manage more complex hazards. By understanding and mitigating the risks, shipping is working to find practical solutions.

Ammonia has been identified as a candidate alternative fuel for shipping due to several characteristics when compared to fuel oil:

- **No carbon**. Ammonia does not emit any carbon dioxide when used to fuel an internal combustion engine, creating the potential for truly zero carbon propulsion.
- **Acceptable energy density**. Although the energy density of ammonia (by volume) is lower than for current marine fuel, it is sufficient to make onboard storage economically feasible.
- **Relatively easy to handle**. Ammonia requires less cooling than hydrogen, for example. It can be stored as a refrigerated liquid at temperatures of around -33°C in relatively simple, inexpensive tanks; or as a pressurised, liquefied gas at ambient temperature and at its saturated vapour pressure. Semi-refrigerated storage under conditions between these two end-points is also possible.
- **Economics have long term potential**. Whilst 'green' ammonia produced from renewable energy is currently costly, prices should fall substantially as production is scaled up.

However, serious safety concerns exist around the use of ammonia as a fuel owing to its hazardous properties, particularly its toxicity.

1.3 PROPERTIES OF AMMONIA

Ammonia is a gas at atmospheric conditions, highly toxic, and corrosive. It is also a flammable gas, although it has a narrower flammable range and higher minimum ignition energy than other fuels such as natural gas and hydrogen, meaning that it is less likely to ignite. The main safety hazard in using ammonia as a marine fuel is exposure through inhalation or skin contact. Relevant properties of ammonia are summarised in Table 1, alongside those of methane (the major constituent of natural gas) as a comparison.

Table 1: Properties of Ammonia and Methane

Property (Units)	Ammonia Value [Reference]	Methane Value [Reference]
Odour (-)	Pungent, suffocating odour [1]	None [1] (unless 'stenching agent' added)
Boiling point (°C)	-33.35 [1]	-161.5 [1]
Autoignition temperature (°C)	651 [1]	537 [1]
Lower Explosive Limit in air (volume %)	15 [1]	5.3 [1]
Upper Explosive Limit in air (volume %)	28 [1]	14 [1]
Minimum Ignition Energy (mJ)	680 [2]	0.3 [3]
Acute Exposure Guideline Level 1 (AEGL-1), 30-minute exposure (nondisabling)	30 [4]	n/a
Acute Exposure Guideline Level 2 (AEGL-2), 30-minute exposure (disabling)	220 [4]	n/a
Acute Exposure Guideline Level 3 (AEGL-3), 30-minute exposure (lethal)	1,600 [4]	n/a

Notes to Table 1:
AEGL-1: The airborne concentration of a substance above which it is predicted that the general population, including susceptible individuals, could experience noticeable discomfort, irritation, or certain asymptomatic nonsensory effects. However, the effects are not disabling and are transient and reversible upon cessation of exposure.
AEGL-2: The airborne concentration of a substance above which it is predicted that the general population, including susceptible individuals, could experience irreversible or other serious, long-lasting adverse health effects or an impaired ability to escape.
AEGL-3: The airborne concentration of a substance above which it is predicted that the general population, including susceptible individuals, could experience life-threatening health effects or death.

1.4 RISK ASSESSMENT

Risk is a combination of consequence (the severity or scale of the consequences of an undesired event) and likelihood (the probability or frequency that the undesired event will occur and result in the consequences).

Risk assessment seeks to arrive at a measure of risk by estimating both the consequence and the likelihood. In many cases these estimates are qualitative, making use of systems that rank likelihood and consequence on scales (such as Low / Medium / High). This is often encountered in the maritime sector, when risk assessment is applied as part of a process to satisfy Class and Statutory requirements for vessel designs that deviate from prescriptive requirements, designs that are considered new or novel, or designs to satisfy goal-based rules or instruments [5].

Quantitative risk assessment (QRA) uses more detailed analysis to calculate numerical values of both consequence (usually occurrence of a fatality or a number of fatalities) and likelihood (usually expressed as a frequency – 1 in a number of years). A form of QRA is outlined in the IMO Formal Safety Assessment (FSA) guidelines [6], together with criteria for risk acceptability. As stated in [6]:

"FSA can be used as a tool to help in the evaluation of new regulations for maritime safety and protection of the marine environment or in making a comparison between existing and possibly improved regulations..."

The benefits of QRA are that it provides a much more granular understanding of risk (which is useful for determining ways in which the risk can be reduced) and enables comparison with numerical risk criteria or benchmarks. However, conducting a QRA is time-consuming and requires a higher degree of expertise.

Although the application of QRA to ships is relatively novel, it has been applied successfully in other industries such as oil & gas [7] and onshore chemicals [8] for many years.

2. OUTLINE OF QUANTITATIVE RISK ASSESSMENT METHODOLOGY

The main steps in the QRA process are illustrated in Figure 1. Each of these steps is described further below.

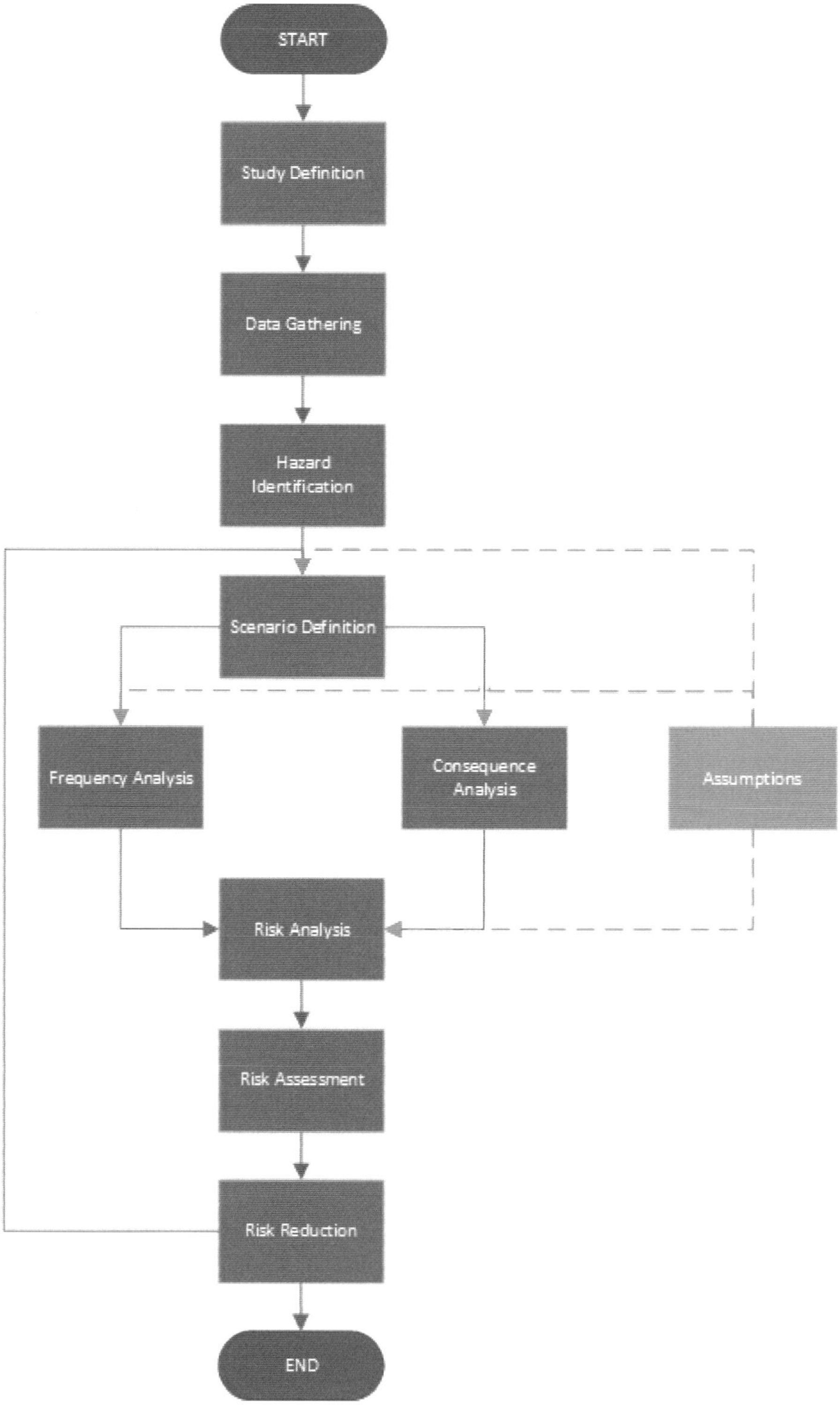

Figure 1: QRA Flowchart

2.1 STUDY DEFINITION

It is important to define the scope and objectives of the study at the outset, since these will influence the breadth, depth and the output of the analysis.

It was agreed amongst the project partners that, at least for Phase 1, the study should be limited to assessment of fatalities amongst personnel onboard. Third party fatalities (such as those amongst shore side personnel or members of the public) were excluded.
Also excluded were releases from vents or ventilation exhausts. These will be the subject of detailed analysis during the next Phase of the safety project.

The reference designs relate to ammonia-fuelled vessels, where ammonia is used as primary fuel, whenever the main engine is at or above a required minimum load. Low sulphur fuel oil (LSFO) or marine gas oil (MGO) is used for the remainder and as a pilot fuel. The studies include the risks from both ammonia and oil-based fuel.

2.2 DATA GATHERING

In terms of data gathering, the reference designs were created by the partners and shared with LRMDH. The data provided included:

- General Arrangement drawings;
- Fuel system Process Flow Diagrams and / or Piping and Instrumentation Drawings;
- Fuel system process conditions (temperatures, pressures, flowrates);
- Details of safety systems;
- Location and dimensions of key spaces on board (such as the Tank Connection Space and Fuel Preparation Room);
- Crew numbers and distribution on board; and
- Typical time spent by the vessel in different operating modes (approaching / leaving berth, mooring / unmooring, moored / anchored, cargo operations, bunkering operations, at sea and laid up / repairing / in dry dock).

2.3 HAZARD IDENTIFICATION

The principal hazards considered were leaks of fuel giving rise to:

- Toxic exposure, fire or explosion in the case of ammonia;
- Fires in the case of HFO and MGO.

The potential causes of leaks were not explored in detail.

The focus was therefore on identifying potential sources of leakage from the components of the fuel system (pipes, pumps, valves, flanges, tanks etc.). This was supplemented by consideration of the possibility of impact events involving the vessel (such as collision or grounding) giving rise to leakage from the fuel tank.

2.4 SCENARIO DEFINITION

The first step in definition of scenarios was to sub-divide the fuel system into 'nodes', on the basis of the reference design information. Nodes are groups of connected equipment items (pipes, tanks, etc.) containing fuel at similar conditions of temperature and pressure. Node boundaries may also be defined by isolation points (such as emergency shutdown valves).

Relevant information on each node was then collated, including the relevant temperature, pressure and normal flowrate, together with the pipe diameters and approximate lengths (where relevant).

2.5 ASSUMPTIONS

The assumptions, covering the other data inputs required by the QRA studies, were compiled into an Assumptions Register by LRMDH and circulated to the other Phase 1 partners for review.

The assumptions were divided into three main categories:

- Operational (O): These are assumptions related directly to the operation of the vessel. For example: manning level, frequency of bunkering, etc. This category was sub-divided by vessel type.
- Technical (T): These are assumptions pertaining to technical aspects of the design. For example: the air change rate in spaces with ammonia-containing equipment, etc.
- Analytical (A): These are assumptions made during the modelling process.

Examples of Analytical assumptions include:

- Criteria for the effect of fire, explosion and toxic exposure on people;
- Ignition probabilities;
- Leak size and duration;
- Leak frequency (likelihood) data used; and,
- Failure probabilities for key items (such as gas detection or isolation valves).

2.6 CONSEQUENCE ANALYSIS

The consequence analysis has been performed using the DNV PHAST™ software, version 8.4.165. The PHAST™ software is internationally recognised as one of the 'industry standard' packages for this purpose. The programme automatically performs all of the required source term, physical effects and impact modelling calculations for each scenario defined by the user. The inputs required are those gathered at the Scenario Definition stage, together with parameters drawn from the Assumptions Register (such as leak duration). Model outputs used were:

- Toxic gas concentration and dose as a function of distance, both for people in the open and those 'indoors' (such as those within the Accommodation);
- Extent of flammable clouds for leaks in the open;
- Thermal radiation flux (i.e. radiated heat) from fires as a function of distance; and,
- Gas build-up in enclosed spaces (using the software's 'In-Building' option).

2.7 FREQUENCY ANALYSIS

The frequency of leaks from equipment has been determined by parts counting (i.e. counting of numbers of each equipment item such as flanged joints, valves, instrument connections, pumps etc.) and application of generic frequency data. The parts count was performed using the reference design information.

The principal source of frequency values was the UK Health and Safety Executive (HSE) Hydrocarbon Release Database (HCRD) and data published by the International Association of Oil & Gas Producers (IOGP) [9]. Where the HCRD data are limited, reference has also been made to the HSE Failure Rate and Event Data (FRED) [10]. These data sources have also been used for frequency analysis of vessels fuelled by liquefied natural gas (LNG) [11].

2.7(a) Event Tree Analysis

A range of outcomes are possible once a leak has occurred, depending on factors such as:

- Whether or not the leak is detected;
- Whether or not isolation is successful;
- Whether or not the leak ignites and, if it does, whether ignition is immediate or delayed; and,
- For leaks in enclosed spaces, whether or not the boundary of the space fails.

These factors are taken into account and the frequency of the possible outcomes calculated using event trees. The principal event tree used is displayed in Figure 2.

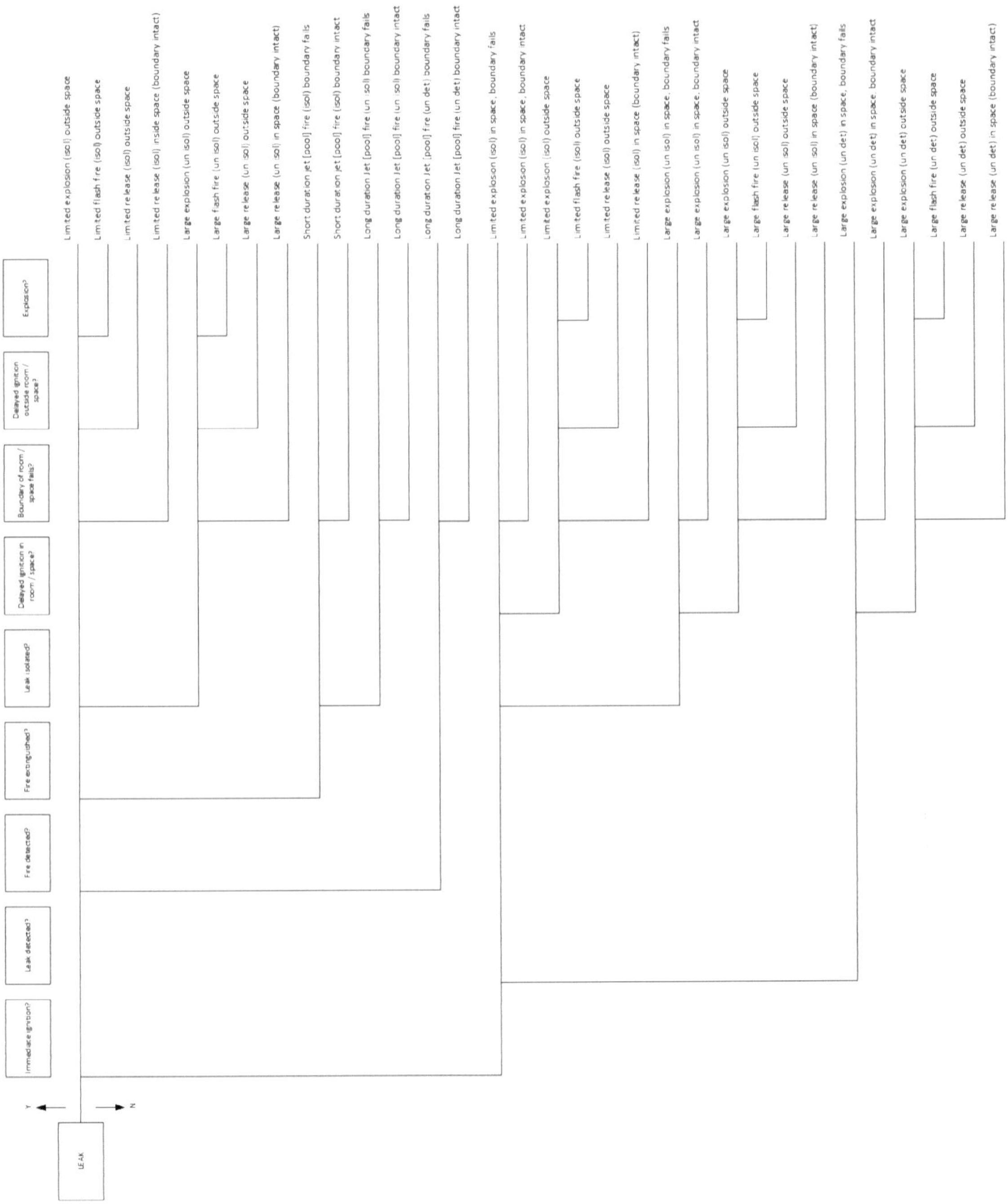

Figure 2: Event Tree

2.8 RISK ANALYSIS

The following measures of risk have been calculated:

- **Location Specific Individual Risk** (LSIR): Risk to a hypothetical individual with defined characteristics permanently occupying a given location onboard. It is used to show which locations have the highest risk, regardless of how much time is spent there.
- **Individual Risk Per Annum** (IRPA): Risk to an individual in a specified group (bridge team, deck crew, engineering crew, etc), who may move from place to place onboard and may not be present at a given location all of the time.
- **Potential Loss of Life** (PLL): The calculated number of fatalities per year.

2.8(a) Calculation of LSIR

In practise, the LSIR at locations of interest around the vessel is calculated first. The LSIR values are then used to generate the IRPA and PLL values.

As outlined in Section 2.7(a), a range of different outcomes are possible depending on what happens after the leak occurs. The frequency of the different outcomes is calculated using event tree analysis. The LSIR at a given location for a given leak scenario and outcome is then essentially:

$$LSIR_{l,s,o} = F_{s,o} . P_{aff,l,s,o} . P_{fat,s,o} \quad (1)$$

For locations that contain sources of leaks (such as a Fuel Preparation Room), the value of $P_{aff,l,s,o}$ for scenarios and outcomes occurring at that location is unity. When a location may be affected by a leak source elsewhere, $P_{aff,l,s,o}$ is a function of direction and the hazardous range of the outcome. The fatality probability $P_{fat,s,o}$ depends on the type of outcome: whether it is just a leak, a pool or jet fire, a flash fire or an explosion.

The total LSIR at the location is then obtained from:

$$LSIR_{tot,l} = \sum_s \sum_o LSIR_{l,s,o} \quad (2)$$

(Note that the probability of a person escaping from the location is dealt with later, during calculation of IRPA).

2.8(b) Calculation of IRPA

For the purposes of calculating IRPA, the crew is divided into crew groups. Each crew group has a defined pattern of where they spend time on the vessel.

First, the IRPA is calculated at each location where the crew group member may be present:

$$IRPA_{C,l} = Y_{F,C} . L_{F,C,l}\left(LSIR_{tot,l} . P_{fail,esc} + LSIR_{tot,MUS}\left[1 - P_{fail,esc}\right]\right) \quad (3)$$

The total IRPA for the crew group member is obtained by summing the IRPA contributions over all locations:

$$IRPA_{C,tot} = \sum_l IRPA_{C,l} \quad (4)$$

2.8(c) Calculation of PLL

Two different approaches to calculation of PLL have been performed for comparison. The first looks at the 'snapshot' distribution of personnel onboard the vessel. It enables a breakdown of PLL by location to be obtained. The second, simpler, method is based on the IRPA results. The second method is more commonly used in oil & gas QRA studies (and includes consideration of escape). The second method can be used to give a breakdown of PLL in terms of the crew groups. The first method is as follows:

Firstly, PLL contributions at each receptor location are calculated as a function of time period and vessel operating mode:

$$PLL_{TP,l} = \sum_{Mod} TP_F . MOD_{F,TP} . N_{l,TP,Mod} . LSIR_{tot,l} \quad (5)$$

The resulting contributions are then summed over receptor locations and time periods:

$$PLL_{tot} = \sum_{TP} \sum_{l} PLL_{TP,l} \quad (6)$$

The second calculation method is simply:

$$PLL'_{tot} = \sum_{C} IRPA_{C,tot} \cdot N_C \qquad (7)$$

2.9 RISK ASSESSMENT

Once the risk analysis results have been obtained, it is necessary to assess their significance. This involves comparison of the results with criteria. Criteria for IRPA are presented in [6] for both existing vessels and new vessels. The criteria for new vessels have been used and are shown graphically in Figure 3.

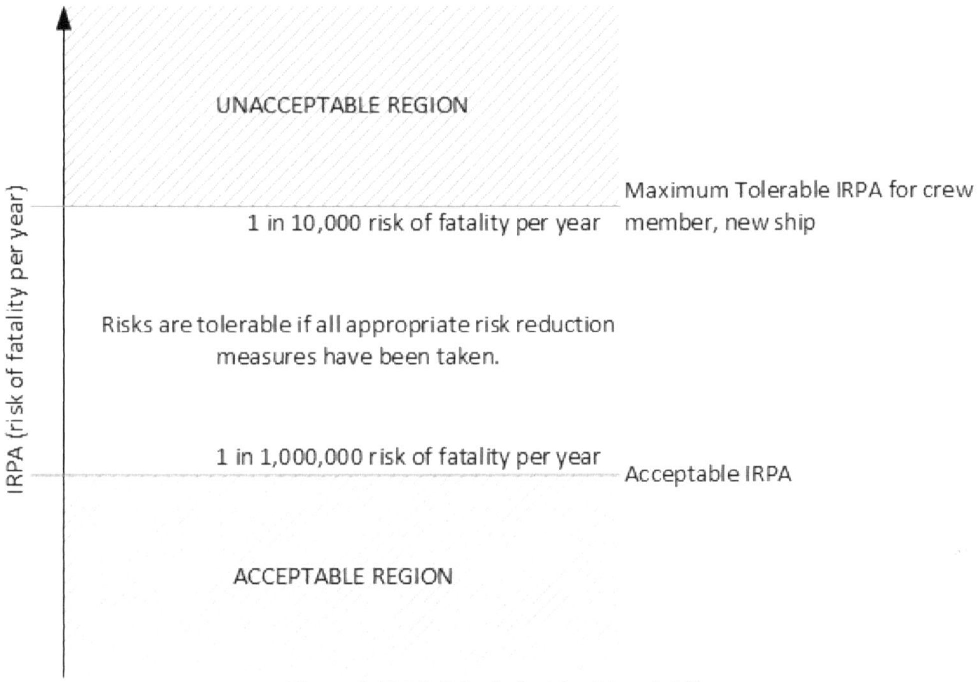

Figure 3: IRPA Criteria for New Vessels [6]

As illustrated in Figure 3, the risk criteria define:

- The level of risk which is deemed unacceptable (except perhaps in extraordinary circumstances); and,
- The level of risk which is considered so low that further efforts to reduce the risk are unnecessary.

Between these two levels is a region in which the risk may be considered tolerable, on condition that all appropriate measures have been taken to reduce the risk.

For the purposes of the project, the judgement about whether an additional risk reduction measure is 'appropriate' will be made using the 'as low as reasonably practicable' (ALARP) principle [12]. The ALARP principle requires that, for risks falling in the middle region in Figure 3, risk reduction measures are applied until the cost of any further measure is 'grossly disproportionate' to (i.e. much greater than) the risk reduction benefit that would be obtained.

Note that criteria are not available for LSIR or PLL. The LSIR values are useful for illustrating the locations of highest risk around the vessel (in the form of a 'heat map') and in calculating IRPA and PLL. The PLL values give an indication of totality or risk for a given vessel, and in judging the effectiveness of risk reduction measures.

2.10 RISK REDUCTION

The risk analysis results may indicate a need to consider the implementation of further measures to reduce the risk. For example, the IRPA results may indicate a relatively high (or even unacceptable) value for members of a particular crew group. Interrogation of the risk results may then be performed to:

- Identify the locations at which the crew group member receives the greatest contribution to IRPA; then,
- At that location, identify the scenarios and outcomes which contribute most to the LSIR.

Options for risk reduction can then be generated, focussing on the highest contributing locations, scenarios and outcomes, for example during a risk reduction workshop. The effectiveness of options can then be tested by adjusting the risk model inputs accordingly and observing the change in risk results. Where multiple options are proposed, they can be tested individually or in combination. This cycle may need to be repeated until a given vessel design arrives at a level of risk that is ALARP.

3. RESULTS

Example, preliminary results are presented to illustrate the way in which QRA can be used. It should be noted that work to refine the designs is ongoing and therefore the results may be subject to change.

The results for an early version of the containership reference design (the 'Base Case') indicated an IRPA for certain crew groups that was close to (but not above) the maximum tolerable level. This was therefore followed by modification of the design to mitigate the risk and re-analysis (the 'Risk Reduction Case'). The results for both cases are presented.

All results shown include the contribution from the periods when the vessel is operating with HFO and not ammonia.

The containership reference design is for a hypothetical 3,500 TEU feeder vessel. Ammonia is stored in a fully refrigerated 1,950 m^3 tank in the hold forward of the Accommodation. For this design it was envisaged that the boiler and auxiliary engines would also be fuelled by ammonia. A Selective Catalytic Reduction (SCR) unit for removal of NOx from the vessel's exhaust was also envisaged, using an ammonia supply. In the first iteration of the design, all of the equipment for supply of ammonia to the main engine, the auxiliary engines, the boiler and the SCR was located in a single large Fuel Preparation Room (FPR), together with the re-liquefaction compressor. The reference design also included the following features:

- Double-walled piping for all ammonia lines outside the FPR and Tank Connection Space (TCS);
- Leak detection triggering automatic closure of isolation valves at the bunker station, as well is in the FPR, TCS and Engine Room;
- Mechanical ventilation of all spaces where ammonia-containing equipment is present at 30 air changes per hour;
- Water curtains at the bunker station, together with a drip tray under the bunker line connections; and,
- Ammonia gas detection in the Accommodation ventilation air intake, causing the ventilation system to switch to recycle.

3.1 BASE CASE RESULTS

The IRPA results for the crew groups onboard the containership base case are shown in Figure 4 (note the logarithmic scale on the y-axis).

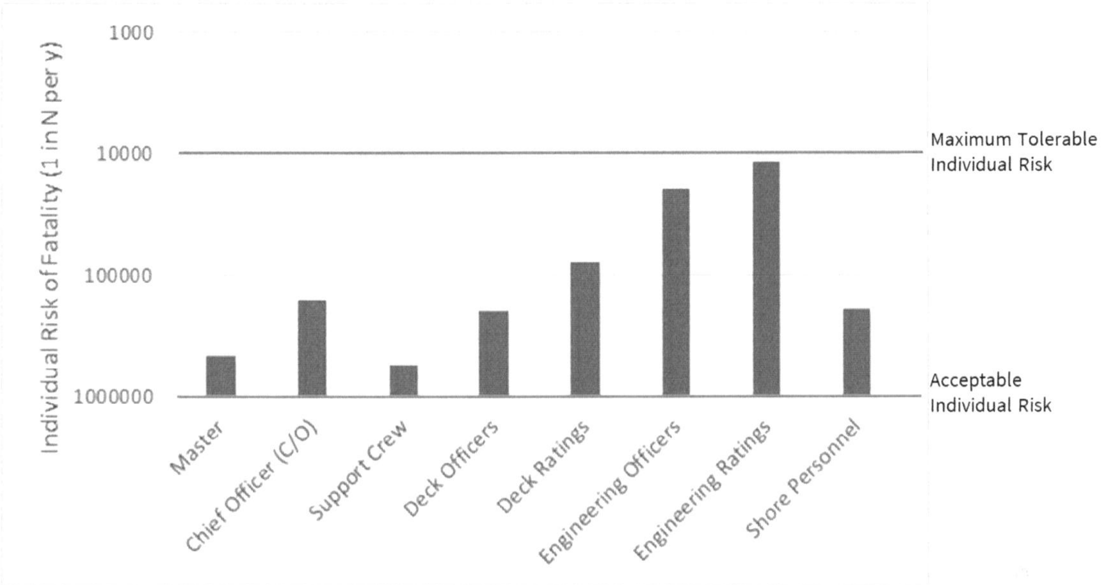

Figure 4: Preliminary Base Case IRPA Results

As can be seen from Figure 4, the crew groups experiencing the highest IRPA are the Engineering Ratings (for whom the IRPA is just below the maximum tolerable level) and the Engineering Officers. The breakdown of the IRPA from ammonia by location for Engineering Ratings is presented in Figure 5.

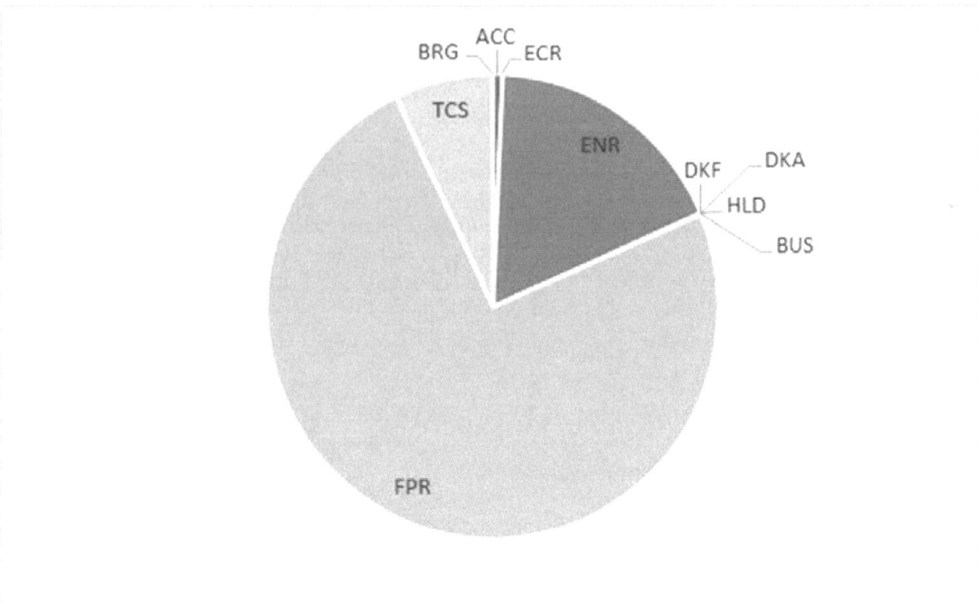

Figure 5: Base Case Breakdown of IRPA by Location for Engineering Ratings

<u>Key to Figure 5</u>:

ACC	Accommodation	BRG	Bridge
BUS	Bunker Station	DKA	Deck Aft of Accommodation
DKF	Deck Forward of Accommodation	ECR	Engineering Control Room
ENR	Engine Room	FPR	Fuel Preparation Room
HLD	Cargo Hold	TCS	Tank Connection Space

Inspection of Figure 5 shows that a high proportion of the ammonia risk for this group arises from time spent in the FPR (even though the proportion of time spent there is very low – around 8-9 minutes per day). It is thought that this is because, whenever entering the FPR, a member of the group is exposed to potential leaks from a large number of sources. It was noted earlier that this space contains equipment supplying fuel to the main engine, equipment supplying fuel to the auxiliary engines, boiler and SCR, and the equipment for re-liquefaction. Therefore, a modification to the design to reduce risk was proposed where the equipment in this space would be split between three separate rooms, one for each of the groups of equipment mentioned. The results for this Risk Reduction Case are presented in Section 3.2.

3.2 RISK REDUCTION CASE RESULTS

The IRPA results for the different crew groups for the Risk Reduction Case are shown in Figure 6, alongside those for the Base Case.

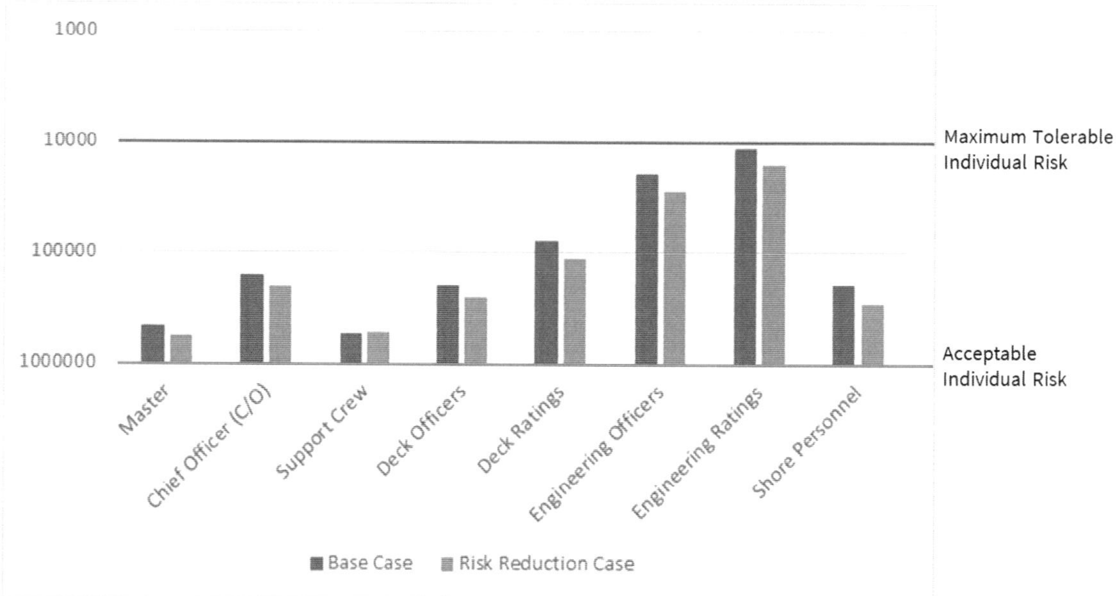

Figure 6: Preliminary Risk Reduction Case IRPA Results

Figure 6 shows that there has been a reduction in IRPA for the Engineering Ratings (as well as some other crew groups). The IRPA for the Engineering Ratings has reduced from 1 in 11,420 risk of fatality per year to 1 in 16,090 risk of fatality per year, a reduction of around 30%. The Engineering Officers' IRPA in the Risk Reduction Case is also further below the 1 in 10,000 risk of fatality per year 'Maximum Tolerable' benchmark.

4. FURTHER WORK

The risk intelligence gained from the QRA modelling in Phase 1 with different vessel designs operating on Ammonia in combination with LSFO / MGO has enable the project team to move into Phase 2 of the project to qualify risk mitigation measures as well as develop human factor related guidelines.

Phase 2 of the Ammonia Safety Project has brought in additional partners: BP Shipping, Cargill, CF Fertilizers, Stolt Tankers, VShips and ABS. The work comprises the following:

- Further risk reduction studies;
- Analysis to test the reference designs and risk mitigations against sets of scenarios derived from HAZID studies and real-world examples of maritime accidents;
- Computerised Fluid Dynamics (CFD) simulation of leaks within an Engine Room (by ABS);
- Finite Element (FE) simulation of collisions between the reference containership and another vessel, to study the potential effect on the ammonia fuel tank (by ABS);
- Analysis of ammonia dispersion from vent masts and from ventilation exhausts;
- A scaling study, to consider how risks would change for larger vessels of the types considered; and,
- A series of tasks relating to human factors, crew training and behaviour.

Phase 2 is scheduled to conclude by the end of 2022.

5. CONCLUSIONS

A detailed quantitative risk assessment model has been developed and successfully applied to designs for ammonia-fuelled vessels. The results may be used for:

- Comparison with published risk criteria;
- Understanding the principal contributors to risks to the crew; and,
- Testing the effectiveness of proposed risk mitigation measures.

For the hypothetical reference containership design described in this paper, the IRPA results were found to be below the Maximum Tolerable level for all crew groups.

It is intended that the findings of the both the Phase 1 QRA studies and the Phase 2 work will be published and used to inform the development of industry guidelines and Class Rules.

6. ACKNOWLEDGEMENTS

In addition to LR and the MMMCZCS, the participants in the study were Maersk, NYK, TotalEnergies, MHI and MAN ES, whose contributions were invaluable.

7. REFERENCES

1. US National Library of Medicine, National Centre for Biotechnology Information Open Chemistry Database Pubchem. https://pubchem.ncbi.nlm.nih.gov/
2. Berufsgenossenschaften, Richtlinien Statische Electrizität, ZHI/200 (1980), Bonn. Buschman CH De Veiligheid 38:20-28 (1962).
3. Babrauskak, V. *Ignition Handbook*, Fire Science Publishers, Issaquah, WA, US (2003).
4. Committee on Acute Exposure Guideline Levels, Committee on Toxicology, National Research Council. *Acute Exposure Guideline Levels for Selected Airborne Chemicals: Volume 6* (2007).
5. Lloyd's Register. *ShipRight Design and Construction. Risk Management. Risk Based Certification* (2021).
6. IMO, *Revised Guidelines for Formal Safety Assessment (FSA) for Use in the IMO Rule-Making Process*. MSC-MEPC.2/Circ.12/Rev 2 (2018).
7. Spouge, J, *A Guide to Quantitative Risk Assessment for Offshore Installations* (1999).
8. Center for Chemical Process Safety, *Guidelines for Chemical Process Quantitative Risk Analysis*. 2nd ed (1999).
9. International Association of Oil & Gas Producers. *Risk Assessment Data Directory – Process Release Frequencies*. Report 434-01 (2019).
10. UK Health & Safety Executive. *Failure Rate and Event Data for use within Risk Assessments. Planning Case Assessment Guide Chapter 6K* (2012).
11. Davies P A and Fort E. *LNG as a Marine Fuel: Likelihood of LNG Releases*. Journal of Marine Engineering and Technology, Vol 12, No 3, (2013).
12. UK Health & Safety Executive. *Reducing Risks, Protection People. HSE's decision making process*. C100. (2001)

8. AUTHORS BIOGRAPHY

Dr Andrew Franks is a Senior Principal Risk Specialist in the Lloyd's Register Maritime Decarbonisation Hub. He has over thirty years' experience in risk assessment and related areas, particularly with respect to hazards with high impact potential. Andrew has conducted numerous quantitative risk assessment (QRA) studies on a wide range of assets handling toxic and flammable materials. He specialises in hazard identification, risk analysis studies and development of risk analysis methods. For the last seven years he has worked on the development of novel risk assessment methods for tankers and other cargo ships, together with development of safety cases for naval vessels. He is currently LR lead on a collaborative project with the Maersk McKinney Moller Center for Zero Carbon Shipping and other partners on the safety of ammonia-fuelled ships.

Claus Graugaard is the Head of Onboard Vessel Solutions at the Mærsk Mc-Kinney Møller Center for Zero Carbon Shipping in Copenhagen, for the past two (2) years and having 25 years of experience in the maritime industry in many technical and managerial global roles. As part of the executive management, he focuses on establishing, strategizing and ramping up the Center and its global activities. In addition to his industry experience, Claus has attended a variety of training programs to develop his technical skills, management & leadership and negotiation competences, giving him vital insights into the international business-oriented aspects of the maritime industry.

AN INVESTIGATION INTO THE POTENTIAL OF THE MOLTEN SALT REACTOR FOR THE PROULSION OF SURFACE SHIPS

B Scott and **Dr T Beard**, BMT, UK

SUMMARY

Molten salt reactors (MSRs) are technically feasible on surface ships, fundamentally, if a safety case can be justified for pressurised water reactors (PWR); the same can be achieved for MSRs, which are inherently safer. While MSR powered commercial ships have challenges associated with security and the regulatory framework to navigate; naval MSR powered vessels must consider crew safety and survivability, with autonomous vessels offering the greatest feasibility.

MSRs are not suitable for all applications of naval or commercial vessel, being limited by vessel principal particulars, the regulatory framework, or survivability. By utilising hydrogen, MSRs or other nuclear reactors can still be incorporated within the vessel energy cycle without needing direct vessel integration; LOHC offers a safe means of hydrogen storage, conjunction with an acceptable energy density. For commercial vessels, the use of LOHC could eliminate regulatory issues associated with emergency planning zones (EPZs), piracy and port access for commercial vessels. For crewed naval vessels, the use of LOHC could guarantee crew safety whilst improving vessel survivability. This paper will further investigate the potential of the MSRs for the propulsion of Surface Ships.

1. INTRODUCTION

In 2018, shipping alone accounted for 2.89% of total global CO_2 emissions [10]; consequently, the industry must take meaningful action that both impacts the near-term (2030), whilst planning for the long-term (2050).

Greater propulsion efficiency and the use of energy saving technologies can reduce greenhouse gas emissions. Whilst they will not meet the targets being set, they should be incorporated to support the uptake of alternative fuels and energy sources, which will be necessary. This is because very few alternative fuels or energy sources have an energy density comparable to Marine Gas Oil (MGO).

The use of alternative fuels such as ammonia, methanol or hydrogen can almost eliminate net vessel emissions. However, these fuels offer significant safety and environmental limitations alongside complex supply chains. For example:

- Ammonia has significant issues around toxicity and NOx, an indirect greenhouse gas, when combusted.
- Methanol requires carbon capture or sourced from biomass to potentially be renewable and would emit CO_2 and NOx when combusted.
- Hydrogen has a significant flammable range and lower minimum ignition energy, as well as being an indirect greenhouse gas if released to the atmosphere.

Without the need for complex supply chains, nuclear power can reliably meet the current and future requirement for low carbon energy whilst maintaining operational capability. The use of nuclear reactors to power vessels using moderately enriched uranium has a successful history, particularly in naval craft. However, these reactors have major environmental, economic, and political implications; especially, concerning decommissioning. These negative considerations could be mostly mitigated by using a MSR. Powered by the radioactive element thorium, a by-product of current mining operations, a Thorium Cycle MSR could offer decreased waste radiotoxicity and significantly improved safety implications when compared with a traditional Uranium reactor.

MSRs have become a buzzword in the energy industry with a huge amount of excitement around the technology and its potential to transform the energy sector. This paper aims to cut through the noise and outline the core technical and operational challenges associated with incorporating a Thorium Cycle MSR into surface vessels for both naval and commercial use.

2. MSR FUNDEMENTALS

MSRs are not a new technology. These reactors were first developed as part of the aircraft reactor experiments conducted by the USA at Oak Ridge National Laboratory (ORNL) in the 1950s. Ultimately, funding was terminated for this project with the safety case for nuclear powered aircraft being unjustifiable; however, the scientists at ORNL were excited by the technology. A second MSR project was conducted at ORNL, with an 8MWt MSR being operational between 1965 and 1969 [17]. Funding for the second project was terminated due to US politics [7].

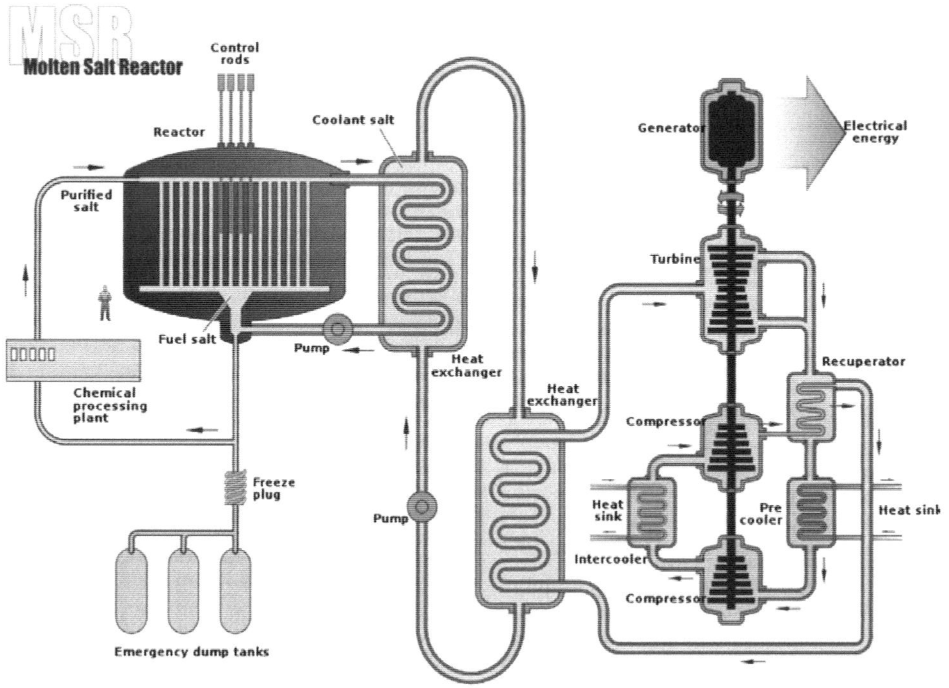

Figure 1: MSR architecture [16].

Figure 1 illustrates typical MSR architecture. This reactor consists of a primary and a secondary molten salt loop. Thorium is suspended within the primary salt loop, the heat generated by the Thorium fuel cycle is transferred to the secondary salt loop which then transfers the heat to a Brayton cycle gas turbine to produce electricity. The secondary salt loop appears to reduce the thermodynamic efficiency of the system; however, this loop reduces the likelihood of the molten salt carrying the radioisotope of Thorium contacting the gas turbine plant due to heat exchanger failure.

MSRs can utilise multiple fuel cycles, the abundancy of Thorium in conjunction with intrinsic proliferation resistance makes this fuel cycle attractive. When compared to Uranium, Thorium is up to 3 times more abundant and most importantly, the waste products of the Thorium fuel cycle have significantly reduced radiotoxicity than the products of the Uranium-Plutonium cycle, facilitated by prevention of the reduced build-up of minor actinides and plutonium [9]. Essentially, the waste products of the Thorium Cycle are radiotoxic for 100s of years as opposed to 10,000s of years [15]; therefore, the waste produced by the Thorium Cycle could be stored in geological facilities without needing to consider long term geological stability.

Unlike PRWs, MSRs are proliferation resistant and operate at ambient pressure, this facilitates passive meltdown prevention when combined with the freeze plug. Given a total failure of the cooling system, the freeze plug will melt, and the primary molten salt will enter the dump tanks. Where the expansion of the salt combined with the nuclear Doppler broadening effect causes the rate of nuclear fission to decrease, allowing the salt to solidify. A paper published by Delft University used a coupled calculation code to model a Thorium MSR, findings suggested that the time taken for the power produced by fission to become negligible in the event of a heat exchanger failure is 3 minutes [8]. Alongside offering improved safety implications when compared with PWRs, MSRs operate at higher temperatures which provides greater thermodynamic efficiency [15].

However, operating with the radioisotope of Thorium suspended in molten salt presents several challenges, the primary technical challenges which currently limit operation are material degradation and tritium removal:

- Material degradation: The reactor vessel, heat exchangers, and pipe work must be resistant to corrosion from molten fluoride salts. This can be mitigated by using INOR-8, an alloy developed by ORNL which has good resistance to corrosion from fluoride salts [1].
- Tritium: Through hydrogen embrittlement, tritium, a radioisotope of hydrogen yields significant operational limitations if allowed to diffuse though the heat exchanges and escape the gas turbine plant. This can be mitigated by using Hitec salt to trap tritium; although, this salt adds complexity and cost requiring three salt loops to prevent the Hitec salt from mixing with the primary salt in the event of a heat exchanger failure [15]. In 2006, ORNL's Charles Forsberg suggested that the use of a Brayton Cycle gas turbine engine can cost effectively facilitate tritium removal, where the tritium is removed from helium [4].

The technical challenges associated with the feasibility of MSRs are not insuperable. With the urgent need to both reduce greenhouse gas emissions and improve energy security, MSRs have attracted significant investment from companies such as: Terrestrial Energy, Filbe Energy, ThorCon, Seaborg, EVOL, MoltexFlex and TMSR-LF. Additionally, the USA's Department of Energy has enabled MIT to lead a university consortium to reassess MSRs. MIT's David Carpenter who's the head of irradiation engineering states:

"As far as next generation reactor technology, this is probably the one closest to being realized. The idea is innovative but not farfetched, and it's not something that will require 30 more years of research." [18].

3. NAVAL ARCHITECTURE

To assess the suitability of utilising a MSR as a vessel's prime mover, the power density, stability, arrangement, and safety must be considered.

3.1 Power Density

Limited information on reactor particulars is currently available, however, a paper published in the Brazilian Journal of Radiation Science compared components of both a MSR and pressurised water reactor (PWR) [5]. Their findings suggest that the equivalent MSR is potentially 60% of the weight, and 80% of the volume of a PWR. Traditional PWRs have already been utilised in surface vessels, perhaps MSRs could facilitate a reduction in principal particulars required to achieve a balanced ship when compared to a PWR powered vessel; although, reactor shielding is likely to still drive required principal particulars.

3.2 Stability

In 2012, BMT's Sir Hill et al investigated the suitability of a MSR for surface ship propulsion [15]. Sir Hill et al discovered that the addition of a MSR, removal of combustion engines and associated fuel, causes the centre of gravity (KG) to increase; however, this is mitigated by no longer being subjected to the free surface effect associated with the fuel oil, resulting in a net increase of ≈4.5% (Figure 2). The increase in overall KG can be offset through achievable changes in loading conditions.

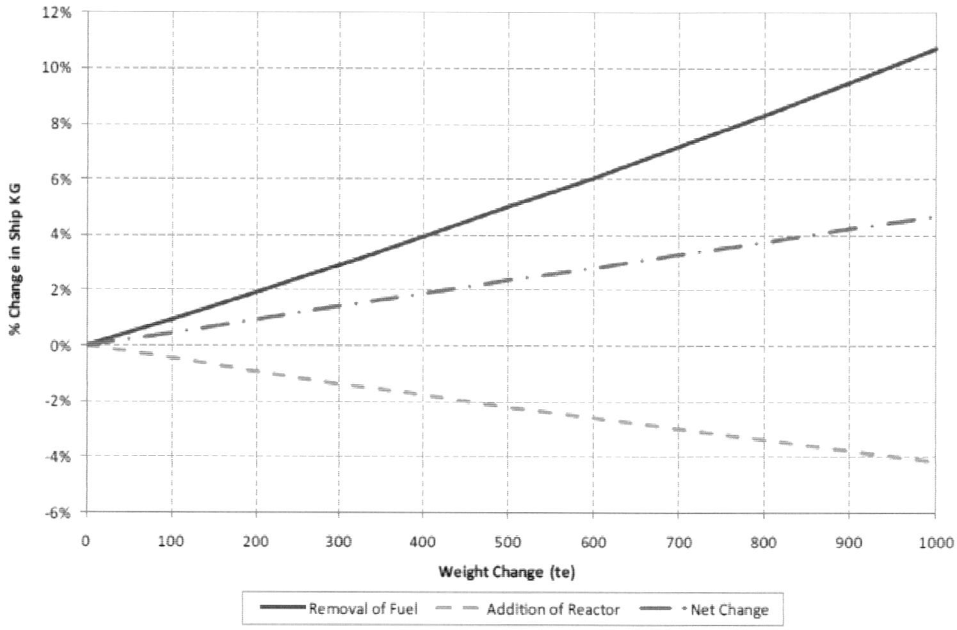

Figure 2: Ship KG vs Weight Change [15].

3.3 Arrangement

Despite having a greater potential power density than a PWR, a MSR would need to be positioned amidships to balance the mass and volume requirements, likely to be driven by shielding requirements. Furthermore, positioning the reactor amidships could minimise the accelerations the plant experiences, thus reduce fatigue; alongside, improving survivability. However, this requirement is likely to make retrofitment challenging and costly.

To further protect the crew from ionising radiation and contain radioisotopes in the event of an emergency, the reactor vessel, emergency dump tanks and Brayton-cycle gas turbine plant would need to be located within the same watertight compartment. Although, the gas turbine plant could be positioned within an adjacent compartment given forecast advancements in tritium capture technology. Attention must be given to the location of the dump tanks, for these tanks must not be in contact with the vessel's outer shell. This requirement may contribute to an increased vertical centre of gravity.

3.4 Safety

Fundamentally, MSRs are inherently safer than PWRs; therefore, if the safety case has previously been justified for PWR powered Naval surface ships, the same can be achieved for MSRs.

The passive walk-away safety, facilitated by the freeze plug and dump tanks primarily contributes to the MSR's enhanced safety case. Given a loss of power or an unexpected increase in reactor power output, the radioisotopes suspended in the molten primary salt will enter the dump tank and solidify; once solid, the rate of nuclear fission will be negligible.

However, utilising the dump tanks will result in the loss of the vessel's prime mover, with a reactor restart requiring significant resources and time. This is a significant concern, both restarting the reactor and maintaining propulsion would require a substantial amount of energy, which would be challenging to either generate or store.

The EPZ must be considered. To maintain the EPZ within the shell of the vessel, such volumes of shielding are likely to result in a reduction of capability for given principal particulars; therefore, MSR powered surface ships will require specialist facilities and insurance when alongside to maintain such a zone.

4. COMMERCIAL VIABILITY

Large displacement commercial surface vessels offer the required principal particulars to achieve a balanced design in terms of weight, volume, and deck area required to accommodate a MSR as the prime mover. Although, after shielding is accounted for the vessel's deadweight will decrease; especially when considering the quantity of shielding required to maintain an EPZ within the ship's boundary, required for operation in densely populated ports. which could be mitigated economically by reduced operational costs. However, the higher initial costs, and potential environmental implications in the event of an emergency will be challenging for insurance companies to quantify and cover accordingly. NS Savannah a nuclear-powered cruise ship, flagged by the United States, was challenging to port; with representatives needing to fly out months ahead to negotiate liability in the event of a nuclear incident [12].

Despite being technically feasible, the regulatory framework poses the most significant challenge for the viability of the nuclear-powered shipping. Initially, classification societies would be required to implement extensive technical studies; IMO conventions, such as MARPOL and SOLAS would need revision to accommodate MSRs. However, the primary challenge is associated with port state controls (PSC).

Figure 3: Location of port state controls.

To facilitate nuclear shipping an agreement is required between PSCs to allow access. A unanimous, global agreement is unlikely; especially when New Zealand's nuclear free zone is considered. A possible solution is 'green corridors', where individual PSCs agree to allow nuclear shipping. Such an agreement between the UK's MCA and the US Coast Guard could allow transatlantic nuclear ships. Such corridors will allow operators and port states to develop ports with designated EPZ; however, maintaining these zones alongside heightened security will incur significant operational costs. Regular naval patrols of these green corridors will be imperative to protect shipping from piracy, with a MSR surface ship being a significant prize.

Nevertheless, regulator progress is being made; in 2022, the UK's MCA drafted to transpose SOLAS Chapter VIII-Nuclear Ships into UK law, being enforceable by the end of 2022 [13].

5. NAVAL VIABILITY

In 2012, BMT assessed the feasibility of the MSR for warship propulsion [15]. Figure 4 illustrates the profile view of the proposed destroyer layout, with the MSR compartment positioned amidships.

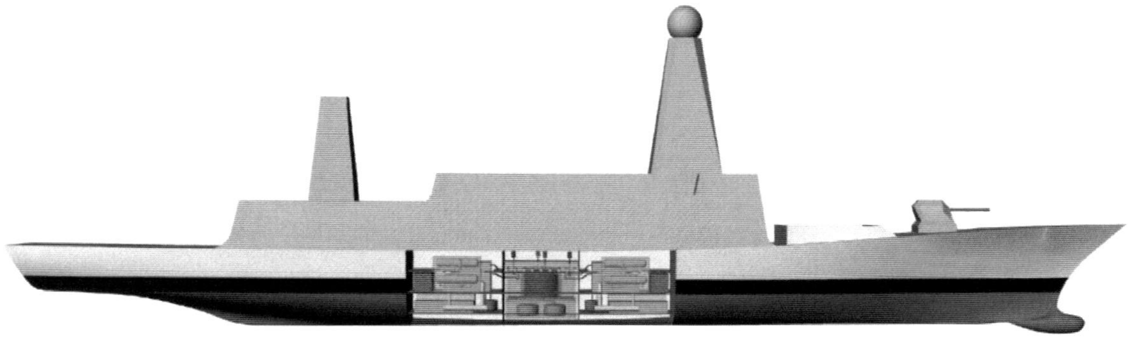

Figure 4: Proposed destroyer layout [15].

BMT concluded that for warship design, significant work is required to quantify thermal and radiological shielding to protect the crew and surrounding environment. Ideally, to improve port availability, the EPZ should be located within the ship; however, such quantity of shielding is likely to reduce current operational capability. For future vessels featuring reduced crew numbers, the volume and mass of shielding could be reduced, further improving the feasibility of smaller warships.

Survivability is the primary concern for warships, particularly combatants. Being exposed to hostile attack, the reactor compartment would require greater protection than commercial vessels require. Despite radioisotopes being rapidly contacted within solidified salt during an emergency, exposure of the contaminated salt is unacceptable for both the crew and the environment. However, for future autonomous warships suitability is improved with the likelihood of crew radiation exposure reduced.

Passive MSR reactor melt down prevention significantly contributes to the enhanced survivability of a MSR powered warship when compared to a PWR powered ship. Although, if activated significant time and power would be required to restart the reactor; needing to transfer the solidified salt from dump tanks back into the primary salt loops. Without a second reactor or backup generator the warship would be floundered, increasing the ships susceptibility to hostile attack.

MSR powered warships are likely to cost less than conventionally, heavy fuel oil powered vessels, even when considering the through life cost. In 2010, the USA's Congress Research Service (CRS) published a report investigating PWR powered surface ships. Their findings suggest that a medium sized surface combatant would have the same total-life cost as a heavy fuel oil powered vessel if the cost of crude oil averaged $70-$225 per barrel [14], at the time of writing crude oil is $85.99 per barrel. Despite the CRS's report being focused on PWR powered naval surface vessel, it is imagined that MSRs would have comparable initial costs yet reduced decommissioning costs.

Whilst the viability of MSR powered surface combatant is limited for combatants due to survivability concerns and space/weight requirements, the feasibility is increased for auxiliary supply vessels which operate further away from the front line. However, a MSR powered axillary ship would be a very attractive target, this is particularly concerning given modern long range high precision missiles and environmental implications in the event of a reactor breach.

Analogous to the commercial application of MSR in surface vessels, the integration of these reactors is not suitable for all surface naval vessels, The fundamental need to guarantee reactor integrity whilst maintaining operational capability limits the use of MSRs to vessels with substantial displacements.

6. ONBOARD ENERGY STORAGE

Clearly, MSRs are not suitable for direct use on all surface ships, however, these reactors can still be integrated into the fuel cycle. This can be done by using the MSR as a power source to transfer energy, either directly as electricity or via alternative fuels.

Using an MSR to recharge electric vessels would potentially have a problem with the power storage. Therefore, the better route is to use MSRs to create hydrogen, a building block for all synthetic fuels. This is supported by using electrolysis of the surrounding sea water, it should be noted that desalination is likely to be required. This poses a problem with the release of the resulting brine, as it could be deemed a local pollutant. However, it is viewed that slight changes in salinity should not be a problem and as such it can be dispersed once mixed with sea water.

The use of hydrogen directly for ships poses a problem due to the energy density, as well as flammability, and as such it is not deemed viable for large-scale deep-sea shipping [2]. Therefore, the hydrogen should be used to create other fuels with greater energy density or fuels that are safer to store.

There is the possibility to use nuclear hydrogen to create methanol, although this requires a supply of carbon dioxide. It would not be viable to extract the CO_2 from the air in the density required and as such would require ships to capture CO_2 and then transfer this when refuelling. This process does have complexity, although research is ongoing to determine the viability of this approach [11]. There is also the issue around the space requirements for the storage, which may not be possible on all vessels.

One of the global favourites for this would be to create ammonia from the hydrogen, ammonia is gaining significant traction as a fuel especially for deep sea shipping. However, ammonia has significant toxicity issues. Meaningful work is being conducted to mitigate the toxic effects of accidental ammonia releases. Ammonia also requires standalone tanks since to keep a liquid, it requires pressure or cooling.

However, there is an alternative which is Liquid Organic Hydrogen Carriers (LOHCs). These are generally aromatic carbon compounds, that can be hydrogenated. For example, a feedstock of toluene when hydrogenated becomes methyl-cyclohexane (MCH). However, the MCH is not utilised onboard a ship and is only used to transport hydrogen. Onboard a ship the MCH undergoes dehydrogenation, producing hydrogen and toluene. The hydrogen can then be utilised in a prime mover, such as ICE or FC. Whilst the toluene can then be stored onboard for transfer. The LOHC cycle is shown in Figure

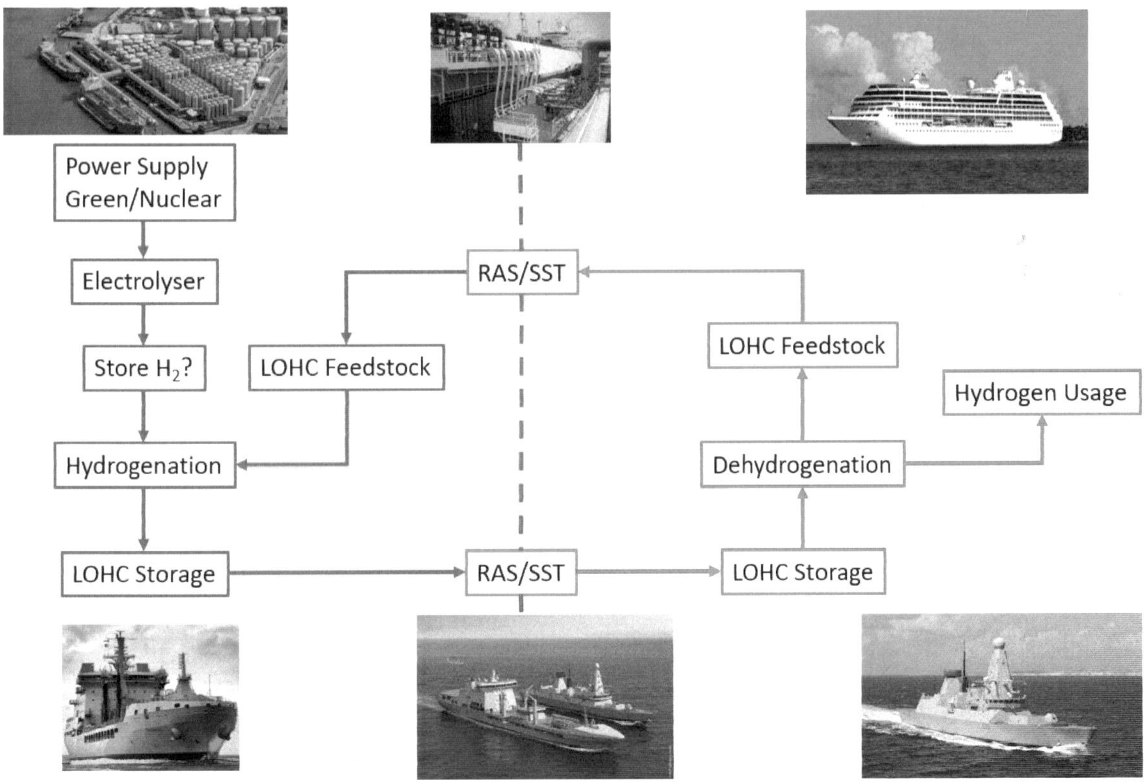

5.

Figure 5: Cyclic nature of LOHC with hydrogen production included.

Both toluene and MCH are liquids at ambient conditions, and as such can utilise existing tanks with some repurposing. This then allows a ship to be designed more efficiently, since there is no longer a need to ballast water as fuel is used. The nature of storing both the hydrogenated and dehydrogenated forms onboard is shown in Figure 6. Whilst the chemicals are toxic since they are stored in a similar manner to diesel and methanol there is a lower risk to crew and passengers. This also means that the amount of hydrogen is limited to what is required for use and is only located in the dehydrogenation equipment, pipework, and prime mover.

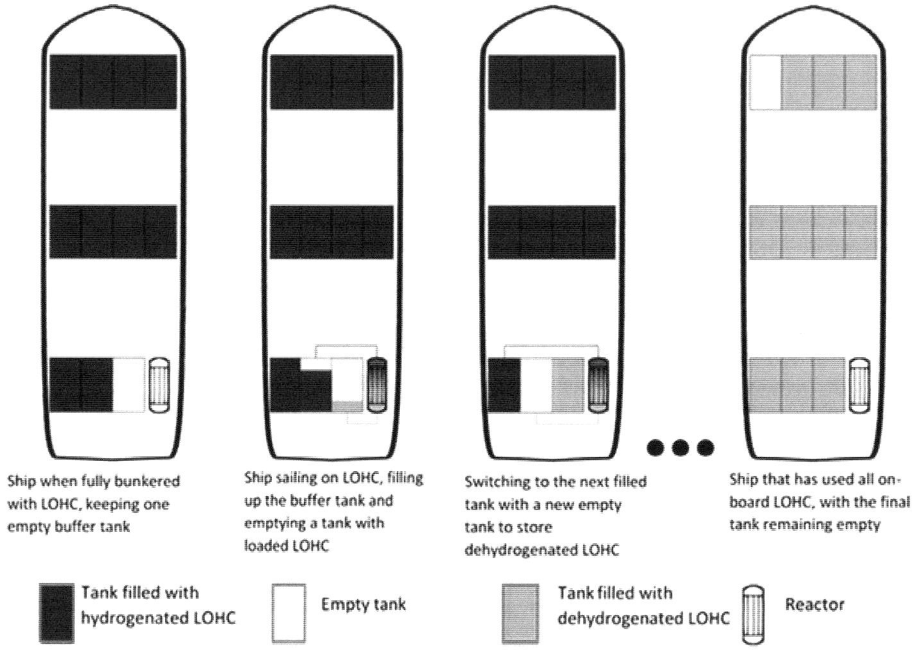

Figure 6: Schematic showing use of a buffer tank for LOHC [6].

Current energy densities for LOHC, based on the MCH-Toluene cycle, are comparable with liquid hydrogen. However, unlike the other fuels, there is far more research possible to increase the energy density by investigating different chemicals. Furthermore, the energy requirement of the equipment required for dehydrogenation reduces the overall energy efficiency. However, this could be reduced by using waste heat recovery with further research potentially offering an energy increase.

7. CONCLUDING REMARKS

MSRs are technically feasible, the challenges of tritium removal and corrosion are not unsurpassable given the current level of investment, alongside the urgent need to generate low carbon sources of energy.

MSRs are theoretically viable on surface ships, fundamentally, if a safety case can be justified for PWR reactors; the same can be achieved for MSRs, which are inherently safer. The primary challenges associated with MSR powered commercial ships are security and regulatory framework; with significant work being required for classification societies, port state controls and insurers. Naval MSR powered surface vessels must further consider crew safety and survivability, with autonomous vessels offering greater feasibility due to reduced shielding requirements.

However, whilst MSR powered surface ships are technically feasible these reactors are not suitable for all commercial or naval surface vessels being limited by vessel principal particulars, regulatory framework, or survivability. By using hydrogen to store energy, MSRs can still be integrated within the vessel energy cycle without being installed on surface ships. LOHCs offering a safe means of hydrogen storage, in conjunction with an acceptable energy density. For commercial vessels, the use of LOHC could eliminate regulatory issues associated with EPZs, piracy and port access. For crewed naval vessels, the use of LOHC could guarantee crew safety whilst improving vessel survivability.

Without the use of LOHC or other synthetic alternative fuel, green corridors combined with regular naval patrols will be imperative for the commercial maritime industry to navigate issues associated with the regulatory framework and security.

In conclusion, there is no ideal alternative energy solution, they all have their benefits and disadvantages which need to be carefully considered for each use case and operational profile. However, driven by rising energy costs and climate change, maritime industry's interest in nuclear energy is increasing. Regardless of the type of nuclear reactor, the use of LOHC or similar synthetic alternative fuel allows for the safe integration of nuclear power. Investment in LOHC powered vessels and infrastructure will achieve an immediate reduction in greenhouse gas emissions, whilst allowing the future integration of nuclear power into the maritime energy cycle.

8. REFERENCES

1. Addison-Wesley, 1958. Chapter 13: Construction Materials for Molten-Salt Reactors. In: Fluid Fuel Reactors. pp. 595-626.
2. DNV, 2022. Maritime Forecast to 2050. [Online] Available at: https://www.dnv.com/maritime/publications/maritime-forecast-2022/index.html [Accessed October 2022].
3. Fordberg, C. W. et al., 2017. Tritium Control and Capture in Salt-Cooled Fission and Fusion Reactors: Status, Challenges, and Path Forward. Nuclear Technology, 197(2), pp. 119-139.
4. Forsberg, C. W., 2006. Molten-Salt-Reactor Technology Gaps. NV USA, ICAPP.
5. Freitas Neto, L. G., Freire, L. O., Santos, A. d. & Andrade, D. A. d., 2021. Potential advantages of molten salt reactor for merchant ship propulsion. Brazilian Journal of Radiation Sciences, Volume 92B.
6. Hoecke, L. V. et al., 2021. Challenges in the use of hydrogen for maritime applications. Royal Society of Chemistry, 14(2), pp. 815-843.
7. Hoglund, B., 2010. Why the molten salt reactor was not developed by the USA. [Online] Available at: http://moltensalt.org/references/static/downloads/pdf/WhyMSRsAbandonedORNLWeinbergsFiringV2.pdf [Accessed 8 October 2022].
8. Hoogmeod, M., 2009. A coupled calculation code system for the thorium molten salt reactor.. Delft: Delft University.
9. International Atomic Agency, 2005. Thorium fuel cycle- Potential benefits and challenges. Vienna: IAEA.
10. International Maritime Organization. (2020). *Fourth IMO Greenhouse Gas Study.* London: IMO.
11. Maersk Mc-Kinney Møller Center, 2022. The role of onboard carbon capture in maritime decarbonization. [Online] Available at: https://www.zerocarbonshipping.com/publications/the-role-of-onboard-carbon-capture-in-maritime-decarbonization/ [Accessed October 2022].
12. Marine Digital, 2022. N.S. Savannah is a nuclear-powered cruise ship. [Online] Available at: https://marine-digital.com/article_nssavannah [Accessed October 2022].
13. Maritime Coast Guard Agency, 2022. Summary of responses on the consultation on the draft Merchant Shipping (nuclear ships) Regulations. [Online] Available at: https://www.gov.uk/government/consultations/consultation-on-the-draft-merchant-shipping-nuclear-ships-regulations-2021/outcome/summary-of-responses-on-the-consultation-on-the-draft-merchant-shipping-nuclear-ships-regulations#next-steps [Accessed October 2022].
14. O'Rourke, R., 2010. Navy Nuclear-Powered Surface Ships: Background, Issues and Options for Congress. s.l.:Congressional Research Service.
15. Sir Hill, R., Hodge, C. G. & Gibbs, T., 2012. The potential of the molten salt reactor for warship propulsion. Bath: BMT Defence and Security.
16. US Department of Energy, 2022. Molten Salt Reactor. [Online] Available at: https://en.wikipedia.org/wiki/Molten_salt_reactor [Accessed 25 April 2022].
17. World Nuclear Association, 2021. Molten SALT Reactors. [Online] Available at: https://world-nuclear.org/information-library/current-and-future-generation/molten-salt-reactors.aspx [Accessed 10 October 2022].

18. Zimmerman, L., 2022. Back to the future: Researchers at MIT's Nuclear Reactor Lab help revive a salt-based nuclear technology abandoned decades ago. [Online] Available at: http://nrl.mit.edu/news/2020/mit-helps-revive-molten-salt-reactors [Accessed 10 October 2022].

9. AUTHORS BIOGRAPHY

Benjamin Scott holds the position of Naval Architect at BMT. He is actively involved in investigating the feasibility of alternatively powered naval and commercial surface ships. Previously, Benjamin has discussed the feasibility of molten salt reactors at conferences and internal workshops.

Dr. Tom Beard is the defence lead for alternative fuel integration at BMT, including the integration between commercial and defence sectors. He has a doctorate hydrogen safety and has presented at various conferences and workshops on the safe integration of hydrogen and its derivatives.

ALTERNATIVE SHIPPING FUELS: MODELLING WIND-FARM-TO-WAKE EMISSIONS

CJ McKinlay, SR Turnock, DA Hudson and **P Manias,** University of Southampton, UK

SUMMARY

The need to reduce emissions from shipping is urgent. Potential future fuel candidates include hydrogen and methanol. This study has attempted to draw a fair comparison between these two fuel types by adopting a bottom-up approach to quantify fuel consumption and emissions. A 10,755 nm voyage undertaken by an LNG carrier was used as a case study. Models were developed for a hydrogen fuel cell energy system and a reformed methanol fuel cell energy system. Simulations calculated the fuel requirements and tailpipe emissions for each option. However, as neither hydrogen nor methanol is naturally occurring, the energy required to produce these fuels should also be considered. Three production methods have been modelled: wind turbines with electrolysis; grid supply with electrolysis; steam methane reforming. Thereafter, the total lifecycle emissions for each fuel option have been calculated and compared to the existing vessel. Typically, this is referred to as well-to-wake emissions, but for green fuels, wind-farm-to-wake may be more appropriate. Results showed that switching to methanol reduced tailpipe emissions by a maximum of 8.3% and wind-farm-to-wake emissions by 18.8% but only if the fuel can be produced entirely from renewable energy. A liquid hydrogen fuel cell energy system produced zero wind-farm-to-wake emissions and required 33.3% less renewable energy than methanol.

NOMENCLATURE

CCS	Carbon Capture and Storage	SoC	State of charge (batteries)
e-fuels	Fuels created from electricity	SMR	Steam Methane Reforming
h	Hours	t	Tonnes
HFO	Heavy Fuel Oil	TFDE	Tri-Fuel Diesel Electric
Kts	Knots (1 knot = 1.94 ms^{-1})	CH_3OH	Methanol
kW	Kilowatts	CH_4	Methane
LCA	Life-cycle assessment	CO_2	Carbon Dioxide
LNG	Liquefied Natural Gas	H_2	Hydrogen
MDO	Marine Diesel Oil	N_2O	Nitrous Oxide
Nm	Nautical miles (1 nm = 1.852 km)	NOx	Nitrogen oxides
PEM	Proton Exchange Membrane (fuel cell)		

1. INTRODUCTION

The shipping sector faces an urgent challenge to reduce emissions. The shipbuilding process has long lead times and ship lifecycles can exceed 25 years; therefore it would be irresponsible to design a newbuild ship today that is not capable of delivering zero emission propulsion. Furthermore, there are several key targets for shipping emissions, such as the International Maritime Organisation's (IMO) target to reduce emissions by 40% before 2030 [1], or the UK government's plans for zero emission shipping to be commonplace by globally by 2050 [2]. However, the shipping industry is yet to reach a clear consensus on the best energy system to deliver zero emission power.

It is likely that the pathway toward decarbonisation will involve e-fuels, fuels that are manufactured using electricity. Two e-fuels that have gained a lot of attention in recent years are hydrogen and methanol. These fuels are manufactured and not naturally occurring unlike currently used oil and gas. Generally, most operators in the shipping sector are first involved with the fuel supply only at the bunkering process. However, for future fuels it may be necessary to also consider the emissions caused by the production and treatment of these chemicals to ensure that the assumed onboard emissions (often referred to as "tailpipe emissions") are not offset by increases in emissions upstream.

For hydrogen and methanol, fuel cells can be used to extract energy as an alternative to combustion. Typically, these electrochemical devices can achieve higher efficiency than combustion engines, and therefore may be valuable to reduce the quantities of these alternative fuels that are required to be stored onboard.

This study has employed dynamic modelling of energy systems, this required the development of mathematical models for components such as fuel cells, engines, and batteries. Thereafter, these models have been used to simulate the fuel consumption and emissions at every point in time against power demand profiles. This can help to inform the future fuels debate by delivering reliable numbers for fuel consumption and emissions based on real world shipping data.

1.1 SCOPE

The emission emitting processes for shipping fuels can be separated into four categories: fuel production; transport and distribution; tailpipe emissions; end of life.

Fuel production emissions for fossil fuels include the treatment processes that turn crude oil and natural gas into fuels used by the shipping industry, such as heavy fuel oil (HFO), marine diesel oil (MDO) and liquefied natural gas (LNG).

For hydrogen and methanol, the processes are different as these fuels need to be manufactured. Therefore, it is necessary to calculate the emissions released during the production process. For this, three different energy sources will be modelled: natural gas using steam methane reforming (SMR); grid electricity; renewable electricity.

The main purpose of this study is to compare emissions pathways for different fuel concepts, all these routes will require the transport and distribution of fuels. The assumption has been made that the transport and distribution emissions would be comparable for all concepts, and therefore will not be covered in detail during this study. Collectively, the fuel production emissions, in addition to the transport and distribution emissions, are referred to as well-to-tank emissions, for e-fuels the terms wind-farm-to-tank or grid-to-tank may be more appropriate.

Tailpipe emissions, or tank-to-wake emissions, are emitted from the energy system onboard the ship. These can be calculated for existing vessels based on fuel consumption and emission factors. For the future fuel concepts, the fuel consumption and subsequent emissions will have to be modelled.

Given the limited timeframe to achieve zero emission shipping propulsion, this study has focussed on technology types that are currently commercially available, although they may have had limited usage at sea to date. For example, a direct methanol fuel cell (DMFC) is a fuel cell type that would not require combustion, however the largest known commercially available model of this has a maximum power output of 500 W. This is significantly lower than the power requirements of even small boats, and therefore has not been considered in this study. Instead, options for methanol include a reformer fuel cell (which converts methanol to hydrogen, then feeds a high temperature fuel cell) and combustion (which also requires a pilot fuel). For the same reason, carbon capture and storage (CCS) technology will not be included in this study, either onboard or during production, as it is unproven at scale [3]. For hydrogen, a proton exchange membrane (PEM) fuel cell has been modelled. All fuel cell setups will include some battery capacity as part of a hybrid system to help meet short fluctuations in demand.

Finally, there are other emissions sources that would typically be included for a complete cradle-to-grave analysis, such as the decommissioning of technologies at their end-of-life. However, this study does not aim to be a complete lifecycle assessment (LCA) and is instead a detailed focus into the production and tailpipe emissions for different scenarios. Therefore, these other processes have not been included. Furthermore, it may be the case that the emissions from these processes are similar for each scenario. Additionally, as other sectors (such as the steel industry) also move towards decarbonisation, then these emissions may decrease without active input from the shipping sector.

The combination of well-to-tank and tank-to-wake emissions are referred to as well-to-wake emissions, again however, wind-farm-to-wake or grid-to-wake may be a more appropriate description for e-fuels.

An alternative method of making methanol would be using biomass, referred to as bio-methanol. Some argue that the carbon absorbed by biomass throughout its growth cycle can offset the tailpipe emissions. However, it is debateable whether biofuels are in fact sustainable. Solomon [4] outlines several key drawbacks to biofuels including scale; efficiency; equity; socio-economic issues; environmental effects and emissions. Additionally, if biofuels do become available, there will likely be scarce supply and the shipping industry would have to compete for supply with other hard-to-abate sectors, such as aviation. For these reasons, this study has focused on e-fuels rather than biofuels.

1.2 AIMS AND OBJECTIVES

The main aim of this study is to compare the well-to-wake or wind-farm-to-wake emissions of different fuel pathways. To achieve this, the following objectives have been outlined:

- Establish a case study voyage from a typical large vessel
- Calculate the emissions and fuel consumption from the existing energy system
- Develop models for fuel cells, engines, and batteries
- Run dynamic modelling simulations to determine the fuel consumption and emissions of e-fuel concepts
- Calculate the emissions caused from several different production options

2. LITERATURE REVIEW

Previous studies have compared alternative fuels for shipping based on real world shipping data [5, 6] however these studies did not use dynamic modelling, instead assuming a constant efficiency for the energy systems. Also, there was a focus on fuel quantities rather than emissions. A similar study used dynamic modelling for shipping energy systems [7] however this study was a comparison of ammonia and hydrogen, rather than methanol, furthermore, supply emissions were again not considered.

This is the first publication to dynamically model a reformed methanol fuel cell for this application. A previous study modelled a high-temperature reformer fuel cell for a cruise ship [8] however the technology was for an LNG-fed fuel cell, and supply pathways were not considered.

Kramel et al [9] developed a well-to-wake model for shipping emissions, however their study focused on global fleet operations, rather than on an individual ship scale. Additionally, Kramel et al [9] considered only current emissions from major fuels (MDO, HFO and LNG) and did not investigate potential future fuel pathways. Ma et al [10] also conducted a well-to-wake analysis of fuels, however this focused only on MDO and HFO. Hwang et al [11] conducted a lifecycle assessment of three fuel options (MDO, LNG and hydrogen) specifically for a coastal ferry, though the only method considered for hydrogen production was steam methane reforming (SMR) that does not have the capacity to produce emission-free hydrogen, unlike some processes such as water electrolysis (using renewable electricity).

Lindstad et al attempted to compare e-fuels to current fuels in terms of well-to-wake emissions [12]. In this paper, Lindstad et al quote zero tank-to-wake CO_2 emissions for e-fuels such as e-LNG, e-methanol and e-diesel, these values have an important impact on the final results, however there is no citation to support this zero-emission assumption. These three fuels all have a carbon content and therefore would release carbon emissions under combustion. The only method to use these e-fuels with zero tank-to-wake emissions would be to employ onboard carbon capture and storage (CCS), this is currently an unproven technology [3] and would have several engineering challenges onboard a ship [13, 14]. Additionally, for a fair comparison then CCS would need to be considered for all fuels including current fossil fuel options. The paper by Lindstad et al also compares fuels on an emissions per unit energy consumed basis, rather than looking at real world ship demand as in this study.

This paper is the first to use dynamic energy system modelling to accurately model the fuel requirements for hydrogen and methanol based on real world shipping data, and to use these results to establish the well-to-wake (or wind-farm-to-wake) emissions for each concept.

3. METHOD

A model has been developed for different energy systems onboard ships, this model adopts a bottom-up approach to simulate the fuel demand and emissions of a specific energy system based on power demand profiles from real world shipping data.

For this case study, an LNG carrier has been used, this vessel completed 66 voyages over a period of 2 years and 5 months, whilst recording several data fields at 30-second intervals. An LNG carrier can be considered a reasonable representation of a typical large-scale international ship as the deadweight tonnage can exceed 100,000 and rated power of 40 MW, which is somewhere between container ships and bulk carriers [15]. This particular ship has 41 MW of rated power. The propulsion system used is tri-fuel diesel electric (TFDE), meaning that either MDO, HFO or LNG can be used to power the generators that produce electricity to power both propulsion and auxiliary systems.

To better understand the fuel requirements of this vessel, a case study voyage was used. This particular voyage was from Singapore to Trinidad and Tobago, it took 25.8 days and represents the largest of all the 66 voyages in terms of: distance travelled (10,755 nm); fuel consumption (2,878 tonnes of MDO equivalent); total energy consumption (15.8 GWh). The route taken by the vessel is shown in Figure 1and the power readings over time are shown later in Figures 4 and 5.

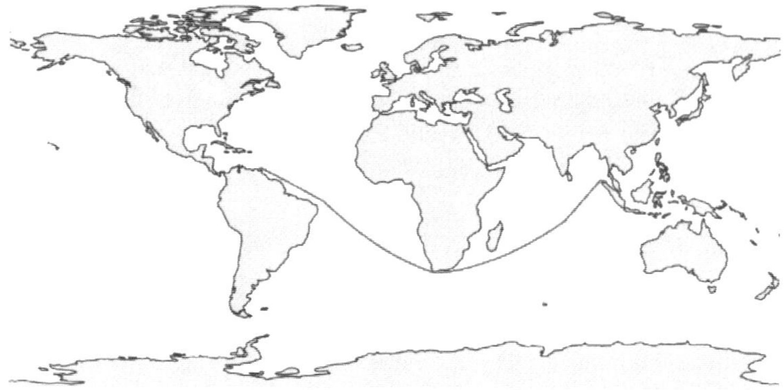

Figure 1. Route for case study voyage.

3.1 EXISTING FUEL SYSTEM

In this section the different fuel scenarios are outlined, including the individual processes that will be modelled throughout the production and deployment.

3.1 (a) Current system

The current energy system on the vessel can run from HFO, MDO or LNG. As this is an LNG carrier, any boiloff from the LNG cargo is directed to the steam turbine generator to act as the main fuel. However, the vessel has the capacity to store both HFO and MDO to also supply power when the boiloff is less than the fuel demand, or when the vessel is undertaking a ballast voyage.

For this scenario, the key processes to model are the post-well treatment of crude oil to make HFO and MDO, as well as the energy required to liquefy the natural gas. In addition, the tank-to-wake emissions will be modelled based on the recorded fuel consumption and emission factor figures. Production and deployment processes are shown in Figure 2.

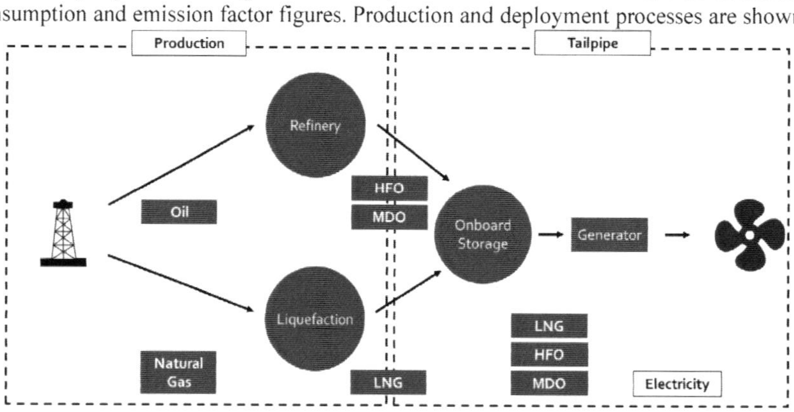

Figure 2. Production and tailpipe emissions pathways for the current TFDE energy system.

3.1 (b) Fuel Consumption and Emissions

Provided data for the case study LNG carrier include recorded fuel consumption at each point in time, the total fuel consumption for each fuel type for this specific voyage are shown in Table 1. From these fuel consumption values, it was possible to calculate the tailpipe emissions based on factors from the IMO's 4th Greenhouse Gas study [15]. This was done for each fuel for the emissions of carbon dioxide (CO_2), methane (CH_4) and nitrous oxide (N_2O), then by multiplying these by the global warming potential factors of 1, 28 and 265 respectively (also from the IMO's 4th Greenhouse Gas study [15]) this gives the total carbon dioxide equivalent emissions (CO_2e) released during the voyage. Furthermore, well-to-tank emissions factors [16] were used to establish the production emissions for each fuel type.

Table 1: Fuel consumption and calculated emissions for the existing TFDE energy system for the case study voyage.

	Recorded Fuel Consumption (t)	Production CO_2e Emissions (t)	Tailpipe CO_2e Emissions (t)	Total CO_2e emissions (t)
HFO	2,139	868	6,765	7,633
MDO	21	12	67	79

LNG	174	67	542	609
Total	2,333	946	7,373	8,320

Results in Table 1 show that the existing fuel system would have produced 7,373 tonnes of CO_2e onboard and 8,320 tonnes from well-to-wake. This will be used as a benchmark for all alternative fuel concepts.

3.2 ALTERNATIVE FUEL PATHWAYS

It is important to clearly define the alternative fuel scenarios and included processes. The fuel transport and distribution emissions have not been included due to the assumption that these would be reasonably comparable for each fuel option.

3.2 (a) Wind-to-Hydrogen
The first alternative fuel pathway to be considered, is a liquid hydrogen system. This will include the use of a fuel cell and battery hybrid system onboard, which can deliver zero emission power. For this scenario, the hydrogen will be produced using renewable energy in the form of wind turbines. Additionally, energy from wind will be used to cool hydrogen to the point where it is liquefied. For this process, there is expected to be zero production or tailpipe emissions. However, there is value in developing this model to understand the onboard fuel storage requirements and the renewable capacity requirements (e.g. number of wind turbines needed). The fuel pathway is shown in the top half of Figure 3.

3.2 (b) Wind-to-Methanol
For a "green methanol" concept, a similar hybrid system will be used. This will have a fuel cell with the same rated power and battery storage with the same capacity, however a Reformed Methanol Fuel Cell will be used rather than a hydrogen PEM fuel cell. For this scenario, there will be some carbon emissions during the reforming process, which converts methanol (CH_3OH) into hydrogen (H_2) onboard.

The production energy will be delivered again from wind turbines, so there should be zero production emissions. The processing will be different though, with the stage of "synthesis" required to produce methanol from both hydrogen and carbon dioxide (CO_2). For this study, it has been assumed that CO_2 would be readily available, however in practice this may not be the case and further energy would be required to produce this molecule. The fuel pathway is shown in the lower half of Figure 3.

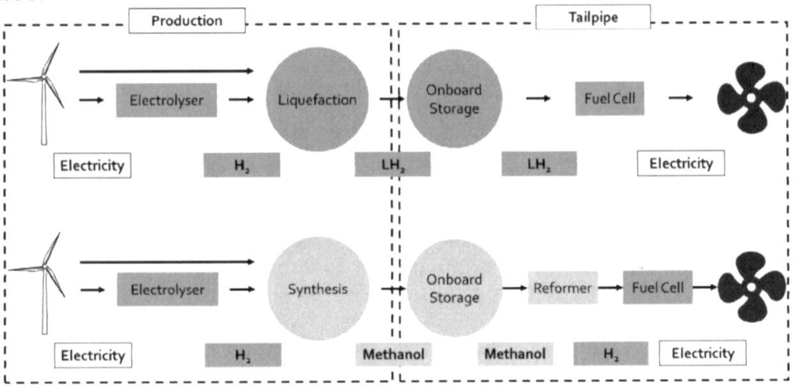

Figure 3. Production and tailpipe emissions pathways for the Wind-to-Hydrogen and Wind-to-Methanol pathways.

3.2 (c) Grid-to-Hydrogen
Unfortunately, renewable energy is not currently an abundant resource, therefore other short term production routes should be considered. The "Grid-to-Hydrogen" concept would be exactly the same as the "Wind-to-Hydrogen" route except that the electricity would be sourced from the national grid, for this a carbon intensity of 232 gCO_2e per kWh of electricity consumption has been assumed (based on the UK grid average in 2021 [17]).

3.2 (d) Grid-to-Methanol
The "Grid-to-Methanol" concept would be exactly the same as the "Wind-to-Methanol" route except that the electricity used to produce the fuel would be sourced from the national grid, for this a carbon intensity of 232 gCO_2e per kWh of electricity consumption has been assumed (based on the UK grid average in 2021 [17]).

3.2 (e) SMR-to-Hydrogen
The previous methods for producing e-fuels used water electrolysis, however this is not currently the most common method of producing these fuels. Instead, the majority of global hydrogen production uses a process called "Steam

Methane Reforming" (SMR) where the hydrogen content of natural gas is extracted. As natural gas contains carbon, there are some emissions released during this process to be modelled. For this case, the hydrogen has been liquefied using grid electricity. An alternative method for producing hydrogen is using SMR with carbon capture and storage (CCS). This is often referred to as "blue hydrogen", however CCS technology is outside the scope of this study. The same onboard hydrogen energy system, as shown in Figure 3, has been used for this scenario.

3.2 (f) SMR-to-Methanol
Methanol is also most commonly produced using SMR, however the process is a little different. Instead of producing pure hydrogen, SMR is used to produce a substance called "syngas" which can be converted into methanol using synthesis. The same onboard methanol energy system, as shown in Figure 3, has been used for this scenario.

3.2 (g) Methanol combustion
A fuel cell is not the only way to extract energy from methanol, instead it can be used in a combustion engine, similar to how ships are commonly powered by existing fuels. This does, however, require a pilot fuel due to the low flashpoint of methanol, in this case MDO has been used. An additional scenario has been included to evaluate the emissions difference between a reformed methanol fuel cell and a methanol combustion engine. This does, however, require a means of initiating the combustion process due to the significantly higher auto-ignition temperature requirement of methanol compared to diesel as a result of the low cetane number [18]. To reach that temperature, compression ratio must be increased significantly (27:1) [18] which requires extensive modification of the engine's internal architecture. Another way that methanol autoignition can be achieved is through the utilisation of a chemical ignition improver or the utilisation of pilot fuel which is typically some form of Diesel [19]. The latter method is selected as it is currently employed in the maritime industry [20].

3.3 DYNAMIC MODELLING

This project uses mathematical models for fuel cells, combustion engines, and batteries, these have been developed from first principles and verified based on manufacturers' specifications. Criteria included in the models are equations based on the ramp rates of each technology and the fuel consumption at partial loads. Of course, batteries do not consume fuel such as a fuel cell or engine would, but the models for lithium-ion batteries do account for ramp rates and round-trip efficiencies.

The value of these models is that they can be used to test different setup configurations against a power profile. For this study, three onboard configurations have been simulated:

- 40 MW hydrogen PEM fuel cells with 10 MWh lithium-ion batteries
- 40 MW reformed methanol fuel cells with 10 MWh lithium-ion batteries
- 41 MW methanol combustion engine

For the fuel cell scenarios, there has been an attempt to run the fuel cells at a constant output for as long as possible, as fuel cells tend to be slower to respond to change than engines. However, it is also important for the state-of-charge of the batteries not to exceed approximately 90% or drop below 20%, as this can increase the degradation rate considerably. To avoid this, the model is set to ramp the fuel cells up or down according when approaching these zones.

The simulations, ran using the programming language Python, calculated the fuel consumption and emissions at every 30 second interval. Hence, delivering reliable outputs for the total tailpipe figures.

3.4 PRODUCTION EMISSIONS

After calculating the onboard fuel consumption, methods were developed to calculate the energy input required to produce these fuels and production emissions.

3.4 (a) Wind Model
The first concepts to consider are the Wind-to-Hydrogen and Wind-to-Methanol scenarios. These processes would use renewable energy both to make hydrogen (through water electrolysis) and to convert the hydrogen into the final product (liquid hydrogen and methanol respectively) so there would be zero expected production emissions. However, there is also value in calculating the renewable energy capacity required to deliver these respective fuels. For this exercise, the annual yield per turbine was calculated based on a typical large 11 MW offshore wind turbine and an assumed capacity factor of 40%, resulting in an annual energy output of 38.5 GWh. This may be slightly conservative as some modern wind turbines can reach capacity factors up to 50% [21].

Gardiner [22] states that the theoretical energy required to liquefy hydrogen is between 3.3 kWh/kg and 3.9 kWh/kg, but in practice the energy is between 7 kWh/kg and 13 kWh/kg. Although, this source is from 2009, so there may have been technological efficiency gains since then. In this case the lower end of the practical quoted range was used, at a reasonably achievable 7 kWh/kg. Including the energy required to produce hydrogen from electricity, which was derived from first principles, the total energy requirement to make 1 kg of liquid hydrogen is 51.8 kWh.

Similarly, the production energy required for methanol synthesis (to convert hydrogen into methanol) was derived from first principles. Including the energy required to produce the hydrogen feedstock, this was calculated as 9.8 kWh per kg. For simplicity, this model has assumed that there would be a readily available supply of CO_2, which is required to produce methanol. However, should this not be the case, then there would be an additional energy cost to produce this.

If the turbine was used solely to produce one of these fuels, the total yield of liquid hydrogen and methanol per turbine per year was calculated as 744 tonnes and 3,933 tonnes respectively. The case study voyage ran for 25.8 days, from entering Singapore port to entering the port in Trinidad and Tobago. Therefore, a better comparison is to consider the quantity of e-fuels that could have been produced per turbine during this period, this equates to 52.6 tonnes for liquid hydrogen or 278 tonnes for methanol.

3.4 (b) National Grid Model

The emissions for the Grid-to-Hydrogen and Grid-to-Methanol scenarios were based off the same production energy values derived in section 3.4 (a) (51.8 kWh/kg for liquid hydrogen and 9.8 kWh/kg for methanol) and a carbon intensity of 232 gCO_2e/kWh [17]. This equates to 12.01 $kgCO_2e$ per kg of hydrogen produced and 2.28 $kgCO_2e$ per kg of methanol produced.

3.4 (c) SMR Model

References from literature were used to determine the carbon emissions from existing SMR production pathways. For hydrogen this was quoted as 10 $kgCO_2e$ per kg produced [23], including the emissions of liquefying the hydrogen using grid electricity (see section 3.4 (b)) this equates to 11.62 $kgCO_2e$ per kg. For methanol production a value of 110 gCO_2e/MJ was taken from the Methanol Institute [24], using a specific energy density of 3.99 kWh/kg (14.4 MJ/kg) this equates to 1.58 $kgCO_2e$ per kg of methanol produced.

4. RESULTS

This study adopted a bottom-up approach, such that the onboard fuel consumption and tailpipe emissions were calculated first, working backwards the production emissions have then been evaluated.

4.1 HYDROGEN FUEL CELL SIMULATION

The results for the simulation based on a setup of 40 MW hydrogen-fed PEM fuel cells and 10 MWh of Lithium-ion batteries are shown in Figure 4, with the latter including the state of charge of the batteries. The energy system was set up to run the fuel cells at a constant output for as long as possible, but they have also been ramped up or down accordingly to avoid overcharging or undercharging the batteries.

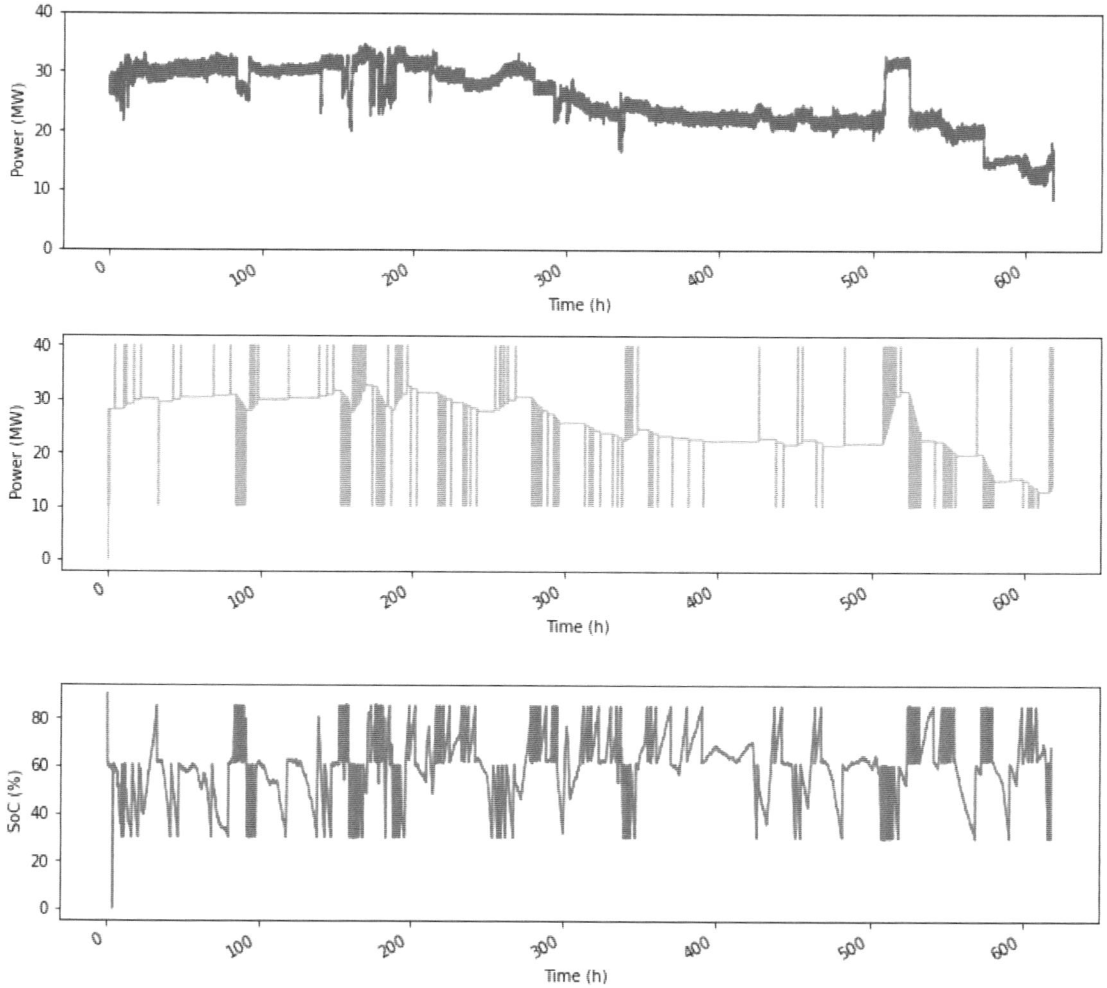

Figure 4. Power profile for case study voyage (upper) with PEM fuel cell output (middle) and battery state of charge (lower).

Simulation results included the hydrogen consumption rate at every 30 second interval, equating to a total of 913 tonnes of liquid hydrogen being consumed during this voyage. As the only by-product of a PEM fuel cell is water, the tailpipe emissions for all scenarios with this energy system are zero.

4.2 REFORMED METHANOL FUEL CELL SIMULATION

A similar simulation has been run for a setup of 40 MW reformed methanol fuel cells with 10 MWh of Lithium-ion batteries, under the same concept of operation, as shown in Figure 4.

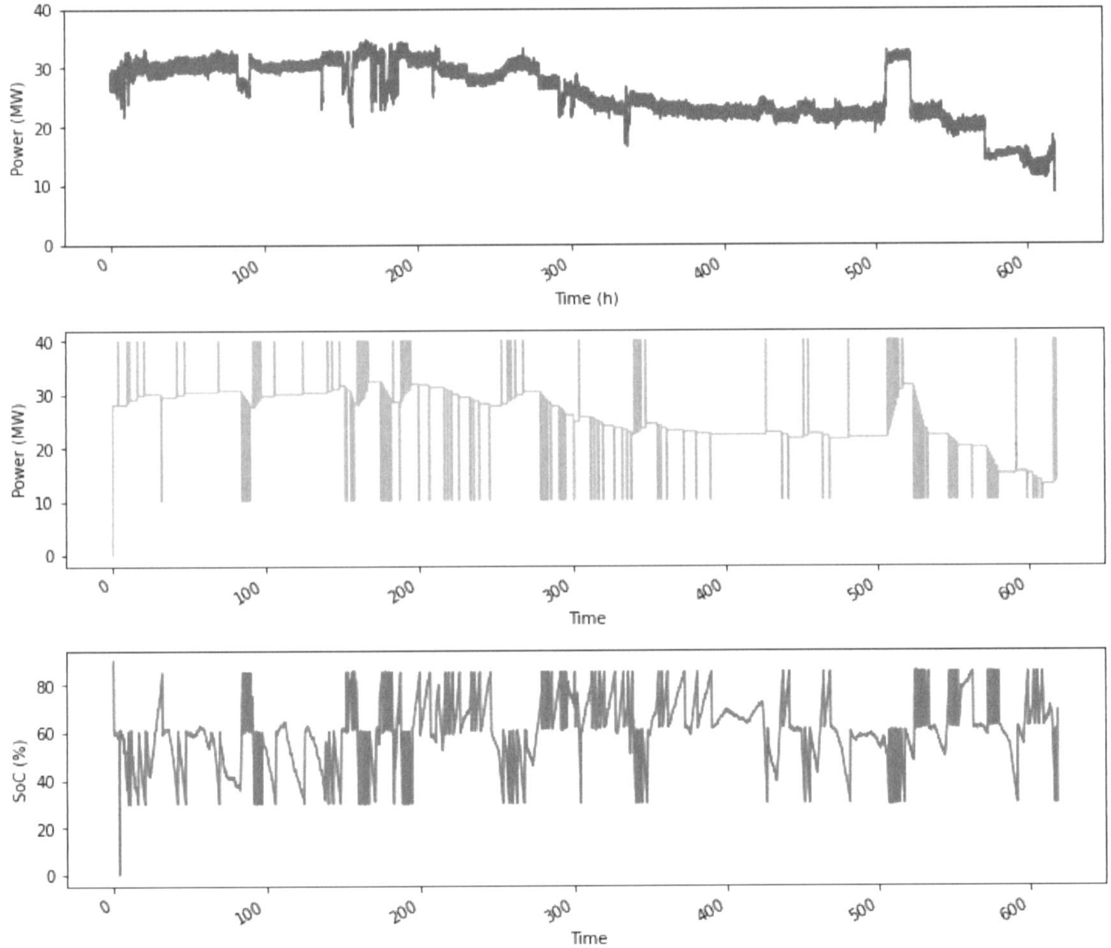

Figure 5: Power profile for case study voyage (upper) with reformed methanol fuel cell output (middle) and battery state of charge (lower).

On this scale the power results appear to be similar to section 4.2, however there are some differences as the high temperature reformed methanol fuel cell will have a slower ramp rate than the PEM. The key result from this simulation was that the methanol consumption would equate to 7,260 tonnes. Also, based on an emissions factor of 0.73 kgCO_2 per kg methanol consumed that was derived from first principles, the total tailpipe emissions would be 6,760 tonnes of CO_2 for this case study voyage. There is not expected to be any methane or nitrous oxides emissions for methanol, therefore the CO_2 emissions and the CO_2e emissions would be the same.

4.3 VOLUME REQUIREMENTS

Currently the case study LNG carrier stores some fuel specifically to provide power for the vessel (MDO and HFO) as well as using some of the cargo as fuel (LNG). For this particular case study, the HFO provided the vast majority of the fuel input and only a small percentage of LNG was consumed, 174 tonnes, which is equivalent to 0.3% of the cargo storage space. It is observable from Figure that almost all the HFO reserve was used, but only a small quantity of the cargo.

Fuel storage requirements are an important consideration for alternative fuels, let's consider if the vessel were to operate entirely on one of liquid hydrogen or methanol. The fuel volume requirements for these fuels have been calculated based on densities of 69.6 kg/m^3 for liquid hydrogen and 1174 kg/m^3 for methanol. By removing the volume of the current HFO and MDO storage capacity, an estimate can be made for the cargo volume that would need to be sacrificed to accommodate these new fuel options. Results show that liquid hydrogen would use 7.7% of cargo space and methanol would use 2.6%. This has been illustrated in Figure 6.

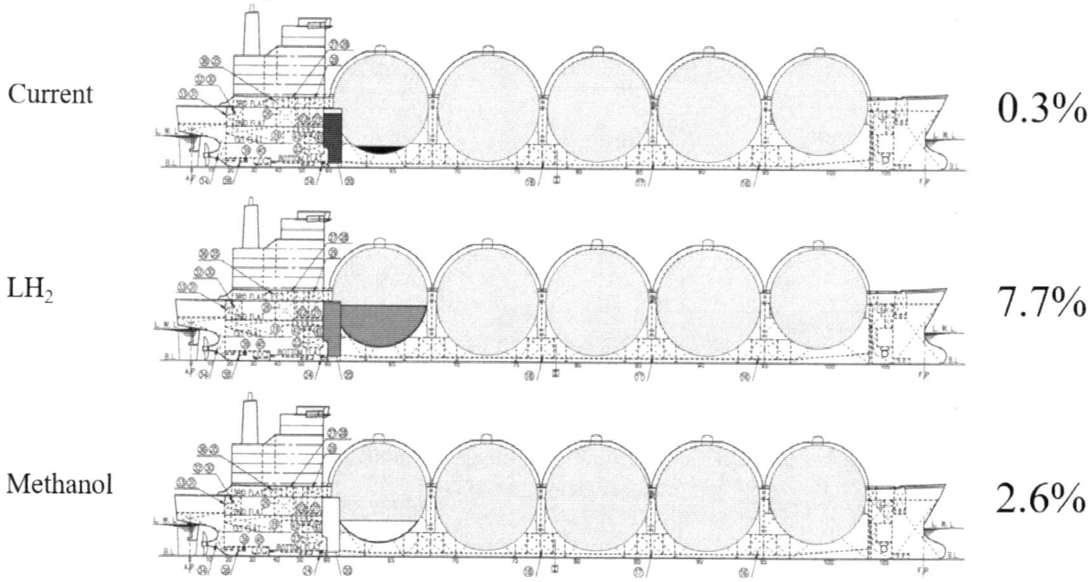

Figure 6: Illustrative fuel storage requirements for the case study voyage including cargo space usage (%). Images are indicative (not actual vessel).

Clearly, it would not be possible to mix fuel storage and cargo as shown in Figure 6, however these images provide a visual comparison. Hydrogen is often dismissed as a future fuel due to its perceived low volumetric energy density, however these levels do not appear to be unrealistic. Furthermore, this voyage took place in 2014, which is before techniques began to be used to reduce energy consumption such as slow steaming. Additionally, there are many further emerging energy saving techniques and devices that could reduce the fuel demand further, such as air lubrication, optimised route planning, and wind assistance technologies. Therefore, the concept of using liquid hydrogen as a shipping fuel should not be dismissed based on the storage volume concerns.

4.4 PRODUCTION EMISSIONS

This section will evaluate the emissions or renewable energy capacity required to deliver the established fuel quantities of either 913 tonnes of liquid hydrogen or 7,260 tonnes of methanol.

4.4 (a) Wind Production Results

Based on the method outlined in section 3.4 (a), the amount of wind turbines required to deliver the fuel for this voyage have been calculated, with the results shown in Table 2.

Table 2: Number of 11 MW wind turbines to deliver e-fuels for the case study voyage over the same period (25.8 days).

Fuel	Energy required to produce (kWh/kg)	Yield (t/turbine/year)	Yield for 25.8 days (t/turbine)	Required fuel for case study voyage (t)	Wind turbines required (count)
LH_2	51.8	744	52.6	913	17.4
Methanol	9.8	3933	278.0	7,260	26.1

As both the Wind-to-Hydrogen and Wind-to-Methanol scenarios are supplied entirely by renewable energy, then there would be zero emissions during production. However, these findings indicate that a methanol supply would require 50% more renewable energy supply (26.1 turbines rather than 17.4 turbines) than the liquid hydrogen scenario. Furthermore, this has not accounted for the CO_2 supply required to make methanol, if this needs to be manufactured then the energy requirements will increase.

4.4 (b) Grid Supply Results

Based on the method in section 3.4 (b), a supply of 913 tonnes of liquid hydrogen would equate to 10,963 tonnes of CO_2e emissions when made from the national grid. Alternatively, a supply of 7,260 tonnes of methanol would equate to 16,521 tonnes of CO_2e emissions when made from the national grid. This would be a significant increase in production emissions;

however most countries are seeing the carbon intensity of their grid power reduce year after year due to the uptake of renewable power. Therefore, this may offer a pathway towards reducing production emissions closer to zero. Although a concern may be the additional load that large scale e-fuel production would put onto the grid.

4.4 (c) SMR production

Based on the method in section 3.4 (c), a supply of 913 tonnes of liquid hydrogen would equate to 10,601 tonnes of CO_2e emissions when made from SMR. Alternatively, a supply of 7,260 tonnes of methanol would equate to 10,681 tonnes of CO_2e emissions when made from SMR. There is less of a distinction here between the hydrogen and methanol production emissions here, which can be attributed to the more efficient step of producing methanol from syngas, rather than hydrogen and carbon dioxide.

4.5 SCENARIO COMPARISON

Now that both the production and tailpipe emissions for all scenarios have been established, these emission values can now be compared, as shown in Table 3.

Table 3: Comparison of well-to-wake and wind-farm-to-wake emissions for each scenario.

Route	Production Emissions (tCO_2e)	Tailpipe Emissions (tCO_2e)	Total Emissions (tCO_2e)	Wind turbines required (count)
Current	946	7,373	8,320	N/A
Wind to Hydrogen	0	0	0	18
Wind to Methanol	0	6,760	6,760	27
Grid to Hydrogen	10,963	0	10,963	N/A
Grid to Methanol	16,521	6,760	23,281	N/A
SMR to Hydrogen	10,601	0	10,601	N/A
SMR to Methanol	10,681	6,760	17,441	N/A

The actual emissions released from the tri-fuel system for this voyage were 7,373 tonnes of CO_2e onboard (tailpipe) and 8,320 tCO_2e including production (well-to-wake). The Wind-to-Methanol scenario results show that the tailpipe emissions would be reduced by 8.3% and the wind-farm-to-wake emissions would be down 18.8%, however there are several criteria required to achieve the latter figure. For example, the methanol would need to be produced from 100% renewable energy and the carbon dioxide required to produce methanol from hydrogen would also have to have zero embodied emissions. Any other production routes are likely to increase well-to-wake emissions, with the Grid-to-Methanol and SMR-to-Methanol scenarios showing significant emission increases of 180% and 110% respectively.

The Wind-to-Hydrogen route does demonstrate a truly zero emission system from Wind-Farm-to-Wake. However, the Grid-to-Hydrogen and SMR-to-Hydrogen scenarios would see an increase in emissions by 32% and 27% respectively. The key difference here, however, the fact that there would be zero tailpipe emissions. Therefore, even if there would be an increase in emissions in the short term, a clear pathway could be established to achieving true Wind-Farm-to-Wake emissions.

4.6 METHANOL COMBUSTION COMPARISON

For methanol, fuel cells are not the only energy conversion technology being considered. Some see methanol as a "drop in" fuel that can be used in a combustion engine like currently used fuels such as MDO. A model has been developed for a methanol combustion engine, which requires an MDO pilot fuel due to the higher auto-ignition temperature requirement of methanol, as described in section 3.2 (g). Results for the case study voyage are shown in Table 4.

Table 4: Methanol combustion emissions comparison for case study voyage.

Route	Production Emissions (tCO_2e)	Tailpipe Emissions (tCO_2e)	Total Emissions (tCO_2e)
Current	946	7,373	8,320
Wind to Methanol (fuel cell)	0	6,760	6,760
Wind to Methanol (combustion)	0	8,381	8,381

It is observable from Table 4 that, even if the methanol could be produced entirely emission free, the emissions actually increased compared to the existing energy system by 13.7 % at tailpipe and 0.7% overall. Therefore, methanol combustion technology should not be considered a viable emission reduction technology for shipping.

5. CONCLUSIONS

Historically the shipping industry has had little involvement in the supply processes of their fuels, with many in the industry hoping that a viable "drop in" fuel will become available in the near future that can reduce emissions with minimal impact on operations. Some present methanol as this drop in fuel, however this study has shown that a change to methanol combustion would actually increase tailpipe emissions by 13.7%. A change to a methanol fuel cell system would reduce tailpipe emissions but only by 8.3%, this is not a sufficient reduction to achieve targets for reducing shipping emissions.

When considering wind-farm-to-wake emissions, a switch to methanol could save a maximum of 18.8%, however this would only be possible if the methanol was to be manufactured entirely from renewable energy. Other methods of producing methanol would increase the carbon footprint of the case study voyage by between 110% and 180%. Therefore, it can be concluded that transitioning to methanol as an alternative shipping fuel is not a viable pathway to meet shipping targets (such as IMO's target of a 40% reduction by 2030).

Another option considered in this study is liquid hydrogen (LH_2) which would be deployed with a PEM fuel cell and battery hybrid. Results showed that 913 tonnes of LH_2 would be sufficient to provide the propulsion and auxiliary power for the longest voyage that the case study vessel undertook in a 2-year 5-month period. It is a common misconception that hydrogen volume requirements are too high for long distance shipping, with this report showing that the equivalent of 7.7% of cargo space would be required to store this fuel. This could be reduced further by employing a range of energy saving techniques and devices.

The liquid hydrogen fuel cell concept is the only route that can deliver zero emission wind-farm-to-wake emissions. However, it would be important to outline a clear pathway to producing the fuel entirely from renewable power, as short-term solutions such as grid supply or SMR would actually increase the overall carbon footprint of the voyage. Additionally, liquid hydrogen was shown to require 33.3% less electricity to produce than methanol (17.4 turbines rather than 26.1 turbines), therefore establishing a pathway to zero emission supply would be a smaller challenge.

6. ACKNOWLEDGEMENTS

The authors gratefully acknowledge financial support from the UK Engineering and Physical Sciences Research Council (EPSRC) through the Centre for Doctoral Training in Energy Storage and its Applications (Grant No. EP/L016818/1).

7. REFERENCES

1. IMO, "Adoption of the initial IMO strategy on reduction of GHG emissions from ships and existing IMO activity related to reducing GHG emissions in the shipping sector," ed: IMO London, 2018.

2. Department for Transport, "Clean Maritime Plan," 2019.

3. M. E. Boot-Handford *et al.*, "Carbon capture and storage update," *Energy & Environmental Science,* vol. 7, no. 1, pp. 130-189, 2014.

4. B. D. Solomon, "Biofuels and sustainability," *Annals of the New York Academy of Sciences,* vol. 1185, no. 1, pp. 119-134, 2010.

5. C. J. McKinlay, S. R. Turnock, and D. A. Hudson, "A Comparison of Hydrogen and Ammonia for Future Long Distance Shipping Fuels," *LNG/LPG and Alternative Fuels by the Royal Institute of Naval Architects,* 30th January 2020 2020.

6. C. J. McKinlay, S. R. Turnock, and D. A. Hudson, "Route to zero emission shipping: hydrogen, ammonia or methanol?," *International Journal of Hydrogen Energy,* vol. 46, pp. 28282 - 28297, 2021, doi: https://doi.org/10.1016/j.ijhydene.2021.06.066.

7. C. J. McKinlay, P. Manias, S. R. Turnock, and D. A. Hudson, "Dynamic modelling of ammonia crackers and hydrogen PEM fuel cells for shipping applications," presented at the International Conference on Computer Applications in Shipbuilding, Yokohama, Japan, 13th-15th September, 2022.

8. P. Manias, C. J. McKinlay, S. R. Turnock, and D. A. Hudson, "Performance optimisation of solid oxide fuel cells through waste heat recovery systems for marine applications," presented at the International Conference on Computer Applications in Shipbuilding, Yokohama, Japan, 13th-15th September, 2022.

9. D. Kramel *et al.*, "Global shipping emissions from a well-to-wake perspective: the MariTEAM model," *Environmental science & technology,* vol. 55, no. 22, pp. 15040-15050, 2021.

10. H. Ma, K. Steernberg, X. Riera-Palou, and N. Tait, "Well-to-wake energy and greenhouse gas analysis of SOX abatement options for the marine industry," *Transportation Research Part D: Transport and Environment,* vol. 17, no. 4, pp. 301-308, 2012.

11. S. S. Hwang *et al.*, "Life cycle assessment of alternative ship fuels for coastal ferry operating in Republic of Korea," *Journal of Marine Science and Engineering,* vol. 8, no. 9, p. 660, 2020.

12. E. Lindstad, B. Lagemann, A. Rialland, G. M. Gamlem, and A. Valland, "Reduction of maritime GHG emissions and the potential role of E-fuels," *Transportation Research Part D: Transport and Environment,* vol. 101, p. 103075, 2021.

13. H. Wang, P. Zhou, and Z. Wang, "Reviews on current carbon emission reduction technologies and projects and their feasibilities on ships," *Journal of Marine Science and Application,* vol. 16, no. 2, pp. 129-136, 2017.

14. T. Smith *et al.*, "CO2 emissions from international shipping. Possible reduction targets and their associated pathways," *UMAS: London, UK,* 2016.

15. Marine Environmental Protection Committee, "Fourth IMO GHG Study," International Maritime Organisation, 2020.

16. DNV GL, "LNG as Ship Fuel, The future-Today," *Hamburg, Germany,* 2014.

17. Electricity Info, "Real Time British Electricity Fuel Mix," ed, 2022.

18. S. Shamun, C. Haşimoğlu, A. Murcak, Ö. Andersson, M. Tunér, and P. Tunestål, "Experimental investigation of methanol compression ignition in a high compression ratio HD engine using a Box-Behnken design," *Fuel,* vol. 209, pp. 624-633, 2017.

19. Y. Dong, O. Kaario, G. Hassan, O. Ranta, M. Larmi, and B. Johansson, "High-pressure direct injection of methanol and pilot diesel: A non-premixed dual-fuel engine concept," *Fuel,* vol. 277, p. 117932, 2020.

20. Wartsila, "Wärtsilä to deliver first dedicated methanol fuel supply system, available at https://www.wartsila.com/media/news/17-03-2022-wartsila-to-deliver-first-dedicated-methanol-fuel-supply-system-3068213," 2022.

21. A. Z. Smith, "UK offshore wind capacity factors Retrieved from https://energynumbers.info/uk-offshore-wind-capacity-factors on 2022-12-07 08:34 GMT," 2022.

22. M. Gardiner, "Energy requirements for hydrogen gas compression and liquefaction as related to vehicle storage needs," *DOE hydrogen and fuel cells program record,* vol. 9013, 2009.

23. T. K. Blank and P. Molly, "Hydrogen's decarbonization impact for industry," *Near-term challenges and long-term potential. Rocky Mountain Institute,* 2020.

24. Methanol Institute, "Carbon Footprint of Methanol," 2022.

8. AUTHORS BIOGRAPHY

Charles McKinlay is a PhD researcher at the University of Southampton, as part of the CDT in Energy Storage and its Applications programme. His research is focused on using dynamic energy system modelling to demonstrate a viable pathway to zero emission shipping. He recently collaborated on 3 Clean Maritime Demonstration Competition projects. His background also includes Mathematics and Energy Engineering.

Professor Stephen Turnock, FRINA is Head of Civil, Maritime and Environmental Engineering at the University of Southampton. He has long standing interests in decarbonisation of shipping and in particular analysis and modelling of alternative power trains including the use of hybrid battery systems.

Professor Dominic Hudson is the Shell Professor of Ship Safety and Efficiency at the University of Southampton. He has research interests in all aspects of ship hydrodynamics, particularly ship design and operation for energy efficiency. He has published over 140 papers in journals and proceedings of refereed conferences. He is a Chartered Engineer and member of RINA. He is presently a member of the 29th ITTC Specialist Committee on Ships in Operation at Sea and the RINA IMO Committee.

Panagiotis Manias is a senior research assistant at the Southampton Marine and Maritime Institute working in the field of marine decarbonisation, through the utilisation of alternative fuels and powertrains with a relevant PhD thesis. He was involved in the powertrain design and subsequent emission modelling for various projects such as a concept of a ballast free liquid hydrogen carrier, and two Clean Maritime projects. His background is in Mechanical Engineering with some industry experience in the technical department of a ship management company.

LOW-CARBON SOCIO-TECHNICAL TRANSITION IN CONTAINER HANDLING EQUIPMENT

C Parro, ESMT Berlin, Germany

SUMMARY

This article aims to investigate, at a strategic level, the drivers in a socio-technical approach to investing in the decarbonisation of the container handling equipment in container terminals. Although there are studies discussing the criteria driving the low-carbon transition, they are not extensive and not focused on the proposed industry.
This study identified the factors and highlighted common patterns, providing helpful information to all stakeholders to create road maps to meet the Paris Agreement goals in the container industry.
Data collection was based on 27 online interviews with experienced professionals. Fifty-nine factors were identified in 7 group categories (political, economic, social, technological, legal, environmental, and internal). A socio-technical structure of the decision-making process was elaborated. The analysis also explored the different leadership styles behind pioneer organizations in adopting zero-emission technologies.

1. INTRODUCTION

The shipping industry corresponds to 2.5-3.1% of the global greenhouse gas (GHG) emissions and, if nothing is done, it could increase between 50% and 250% by 2050, undermining the objectives of the Paris Agreement [1].
While shipping plays a central role in global supply chains, meaning many industries will use the shipping sector to facilitate their net-zero goals [2], the majority of the studies address the problem of the emissions generated by the vessels [3] under the leadership of the International Maritime Organization (IMO), but few attention and, in a more scattered way, has been devoted to the terminals themselves and even their methodology to calculate their carbon footprint [4].
More and more, port authorities and container terminals will face the pressure of end-users, the public opinion, and regulators regarding their sustainability compromise _ accommodating social impacts such as traffic in the cities; waste treatment; health and safety of the employees and nearby population; emissions; noise pollution; and so on.

1.1. BACKGROUND

1.1 (a) The Paris Agreement

In December 2015, the historic legally binding international treaty on climate change was signed in Paris with the goal to limit global warming to 2, preferably to 1.5 degrees Celsius, compared to pre-industrial levels. The 197 signatories aim to achieve a climate-neutral world by mid-century [5].
Zero-carbon solutions are becoming more relevant across various economic sectors that account for 25% of total emissions. This trend is undeniable in the electricity and transportation sections. But one of the critics raised by one of the interviewees of this study is the lack of a clear path of achieving the goals. "Imagine that the world decided to build a house. By 2030, we will have the first floor; by 2050, 3 floors. But nobody knows who does what. What should I bring? Bricks? Paint?", she outlined.

1.1 (b) The 2020 letter to CEOs

In January 2020, in his traditional annual letter to CEOs (Chief Executive Officers) [6], Larry Fink _ the Chairman and CEO of BlackRock _ defended that sustainability and climate-integrated portfolios are the only way to manage future risks and guarantee returns to investors who now have a deeper understanding how climate risk will impact both our physical world and the global system that finances economic growth.
A tremendous pressure is created in companies and policymakers when the world's largest asset manager, with just over US$9 trillion in assets under management, announces several initiatives to place sustainability at the center of their investment approach and it could potentially reshape the entire finance system
While container terminals are not in the spotlight of the decarbonisation acts, building a container terminal is a capital-intensive economic activity. Companies that do not address sustainability risks will encounter in their horizon growing scepticism from the markets, and in turn, a higher cost of capital [6]; this puts the problem at the center of the strategic discussions in the industry while endangering the viability of, not only projects but the activity itself.

1.1 (c) Current Situation of External Policies and Regulations and Future Prospects

Some governments and port authorities have strategies, clear goals, regulations, monitoring, and funding in place, providing a more fertile setup for investments led by terminals.
In California, the Clean Air Action Plan, adopted initially in 2006 and last updated in 2021, is a landmark air quality plan that establishes a comprehensive and far-reaching strategy for reducing port-related air pollution and related health risks while allowing port development [8].
In the Far East, the Port of Hong Kong has an Air Pollution Emission Reduction Plan and Air Quality Objectives in place for at least 10 years [9]. In Oceania, the port of Auckland in Australia committed to zero emissions by 2040 [10].
The major European ports assumed the compromise of being carbon neutral _ Rotterdam [11], Valencia [12], Antwerp [13], Peel Ports in the UK [14]. The list is not extensive and will keep growing in the following months.
For the future, a price increase in diesel is expected due to detaxation and/or taxation, and carbon taxes may be implemented [15]. Additional restrictions covering emissions are also forecasted for the licensing of future projects.

1.2. THE IMPORTANCE OF CONTAINER TERMINALS

There are more than 600 container terminals worldwide, with 200 classified as mid-large handling traffic above half a million TEUs (twenty-foot equivalent unit) [16]. In 2020, the total throughput of containers was 815 million TEUs, a decrease of 1.2% compared to 2019. The industry is expected to grow between 4 and 6% in the next 5 years [17].

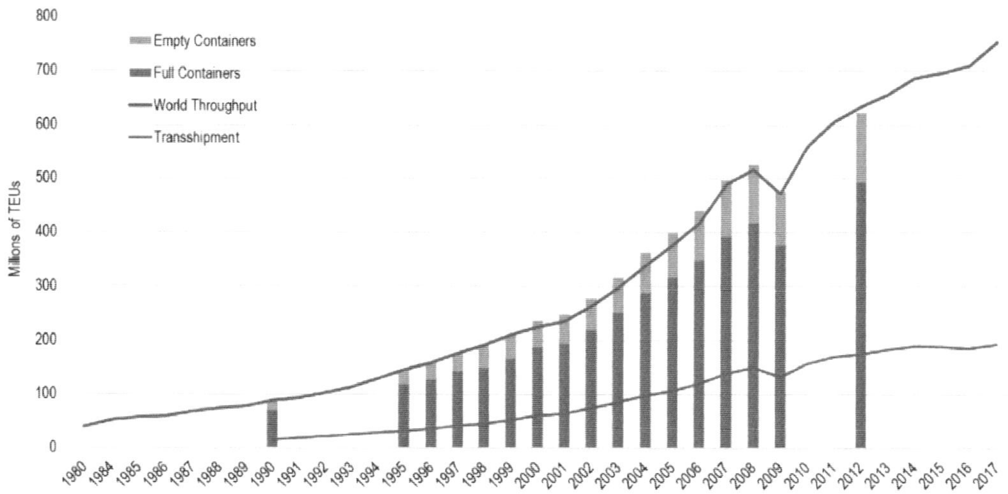

Figure 1: World container throughput 1980-2017 in millions of TEUs [16].

1.3. WHAT ARE CONTAINER TERMINALS DOING?

There are proposed methodologies to calculate the carbon footprint of container terminals [4]; still, the industry lacks a standard and, the result is each port authority, and each terminal operator is creating their own blueprint.
This section will summarize and categorize the principal investments container terminals make in their low-carbon transition.

- Alternative fuels. Biofuels made of crops, or organic rejects have high potential because no modification is necessary for the currently used diesel engines. They will most likely be complemented by synthetic fuels such as LNG, LPG, methanol, ammonia, and hydrogen [18].
- Renewable energy. Terminals are switching to green electricity produced by hydropower, wind, or solar systems when available in their operating location. However, the investment in the generation of green electricity by the terminals is not very popular.
- Technological measures. Monitoring systems and smart controlling solutions, new building techniques and materials, regenerative solutions, hybrid equipment, more efficient powertrains, and electrification are not an extensive list of actions to improve energy efficiency.
- Operational improvement. Such as optimizing routes and utilization of equipment, coordination of external demand such as road trucks and vessels, and training of operators could bring substantial emission reductions.

The investment in environmental projects represents only 6.3% of the total CAPEX of ports [17].

1.4. CONTAINER HANDLING EQUIPMENT

A container terminal is divided into 2 types of operations: quayside and hinterland. For each operation, there are different types of equipment moving the containers from the vessel to the terminal, then transferring them to a yard where yard cranes will stack them. Then, the containers can be picked up by external trucks and leave the terminal, or they can be loaded on a train.

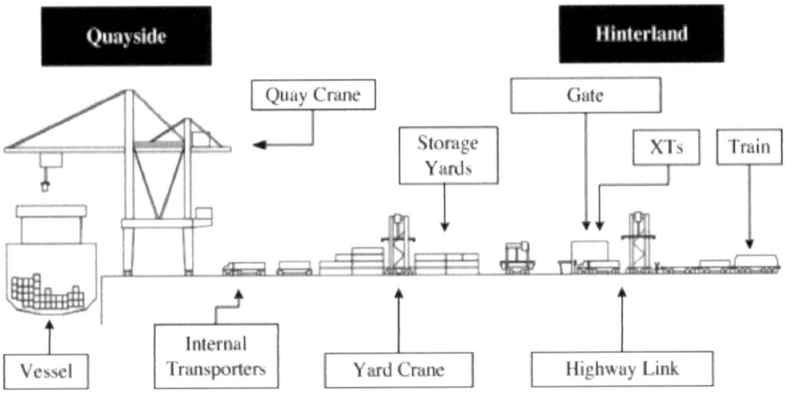

Figure 2: Basic arrangement of a container terminal [19].

This study will focus on initiatives to reduce emissions in container handling equipment (CHE).
Figure 3, produced from a study with 31 terminals in 16 countries [4], shows the energy consumption converted into diesel equivalent, confirming that horizontal transportation and yard equipment, referred to as "horizontal" in the chart, are the highest share. Since electrification in yard cranes is a mature technology with almost 20 years in the market, despite the operational model, the biggest challenge still is on horizontal transportation equipment such as terminal tractors, lift trucks, and straddle/shuttle carriers. This equipment is most commonly diesel-powered. Because of their mobile nature, electrification is not as easy as rubber-tired cranes, and it brings a series of challenges such as the size of batteries, recharging time, and high CAPEX.

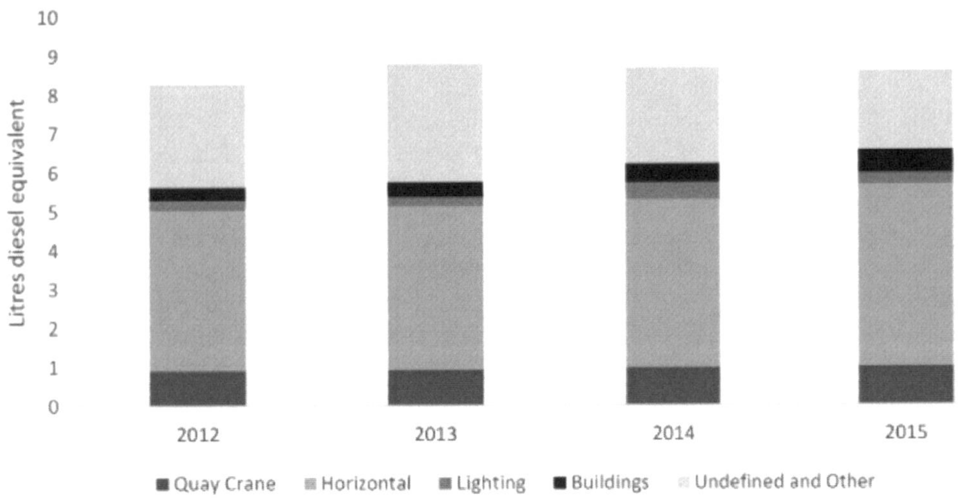

Figure 3: Median litres of diesel equivalent consumed per activity cluster (excluding reefer) in container terminals [4].

1.5. RESEARCH QUESTION

This thesis will investigate the drivers for and the barriers to investing in solutions to reduce GHG emissions in container handling equipment.

The interviews will be applied to industry experts from five niches of the container industry: container handling equipment manufacturers, container terminal operators, sub-suppliers, port authorities, and utilities. Focus is placed on understanding the underlying factors that lead to the successful development and adoption of new technologies.

The results will support boards, suppliers, technology developers, policymakers, and buyers in implementing a more proactive decarbonisation strategy optimizing resources and minimizing risks.

1.6. SCOPE LIMITATIONS

The scope will be limited to deep-sea container terminals and container handling equipment industries.

No further investigation will be done in emissions originating from vessels, buildings, electricity, and other vehicles. No other sustainability aspects will be explored.

This study does not aim to produce a quantitative analysis but to extract, identify and categorize the professionals' insights.

2. LITERATURE REVIEW

A literature review was first conducted to identify theoretical models and, likewise, studies exploring the criteria applicable to the maritime industry to support the understanding of low-carbon transitions and the role of leadership in those transitions.

2.1. SOCIO-TECHNICAL TRANSITIONS: THE MULTI-LEVEL PERSPECTIVE

The Multi-Level Perspective (MLP) on Socio-Technical Transitions such as the low-carbon is an innovative and comprehensive approach that conceptualizes those transitions that arise from the alignment of processes in three analytical levels: niche innovations, socio-technical regimes, and the socio-technical landscape. It addresses the gap in energy studies and modelling research that so often dominates academic and policy debates; even though they provide valuable insights, they remain constrained by their focus on rational decision-making and their neglect of non-linear dynamics and broader social processes [20].

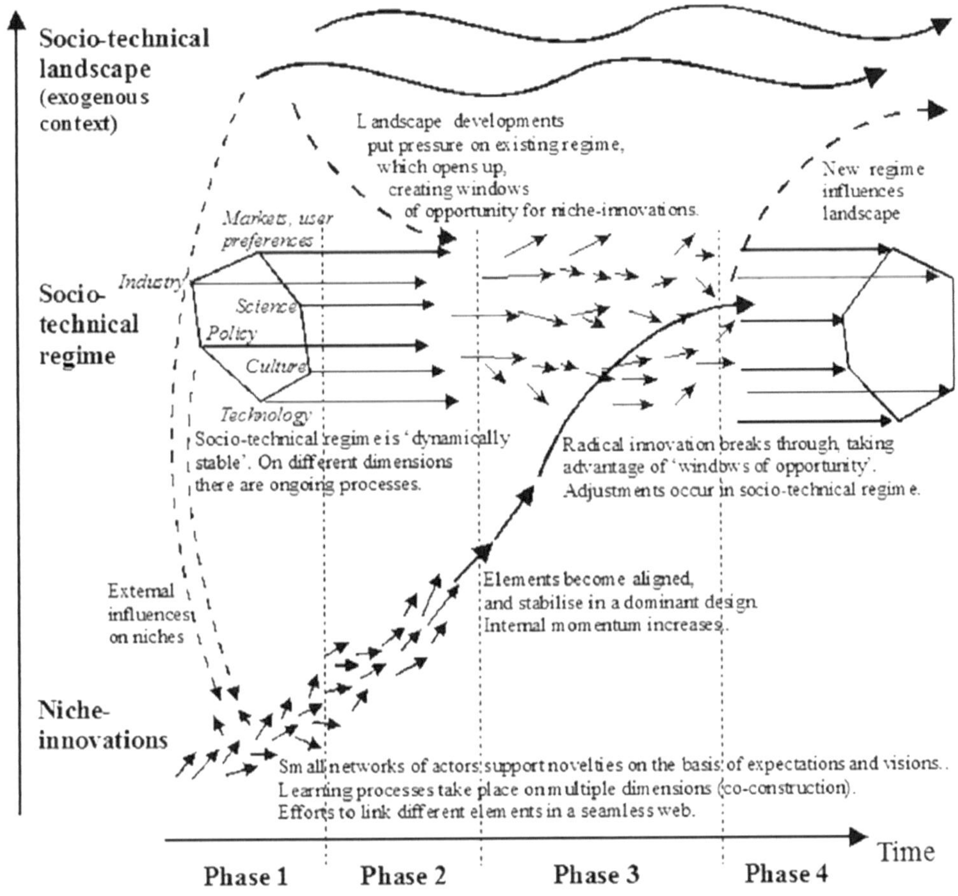

Figure 4: Multi-level perspective on socio-technical transitions [20].

The multi-level perspective can be successfully applied to describe the energy market [21] and to analyse the transition to low-carbon transportation in socio-technical systems [22]. A transition management approach was also derived from the multi-level perspective [23].
Although the Multi-Level Perspective on Socio-Technical Transitions provides an illustrative understanding of how the low-carbon transitions happen, it has limitations explaining the strength of the relationships between different elements [24]. Without understanding the relationships, there is also limited comprehension of the non-linear effects that a new input can cause in the entire system.
The MLP created the grounds for this study since a similar phenomenon was identified in the low-carbon transition in the container handling equipment industry.

2.2. SUSTAINABILITY CRITERIA IDENTIFICATION IN MARITIME INDUSTRY

Several studies cover the harmonization of sustainability criteria focusing on shipping, mainly in alternative marine fuels [25]. Others propose drivers and discuss their potential with more practical suggestions for reducing emissions in vessels operations [26].
On the side of the terminal, macro factors affecting the development of smart green ports are well covered [17] and, a more in-depth sustainability assessment criteria for container seaports can also be found backed up by experts [27][28].
A literature review on the elements which may influence company responsiveness to the implementation of sustainability initiatives covering external and internal factors is also available [29].

2.3. THE LEADERSHIP ROLE IN LOW-CARBON TRANSITIONS

The times have changed, and so have the obligations of leaders, who now have stakeholder capitalism directing their accountability. Making money is no longer sufficient; how the money is made is sometimes scrutinized as well. This poses significant hurdles for effective leadership strategies since it broadens the breadth of expertise needed to manage all aspects of a company.

Studies discuss the new leadership abilities necessary to address the sustainability challenges [30][31].

Other literature focuses on the learning process of preparing sustainability leaders [32][33][34].

Finally, further research looks at the power leaders have in steering the company strategy incorporating sustainability [35] and political leaders influencing the development of sustainable goals [36].

3. METHODOLOGY

Initially, a literature review was conducted to identify previous criteria for sustainability decision-making. Then, a questionary was created, and interviews were conducted to determine the drivers for investing in zero-emissions technologies based on a multi-stakeholder participatory approach.

3.1. INTERVIEWS DEVELOPMENT AND PARTICIPANT RECRUITMENT

The questionary had 10 open questions, and it was planned to be applied to up to 30 subjects in online recorded interviews between January and February 2022. The list of questions can be found in Appendix A.

Subjects' responses were confidential since they gave access to personal experiences, professional opinions, and individual insights and observations that could be sensitive.

Five niches of the container industry were targeted: container terminals, equipment manufacturers, sub-suppliers, utility suppliers, and port authorities. Container terminals refer to professionals (or former professionals) working at container terminals or headquarters managing a portfolio of container terminals. Equipment manufacturers refer to professionals working for companies producing CHE. Sub-suppliers apply to businesses providing components, such as engines, to the equipment manufacturers. Utility suppliers refer to oil or utilities companies. Finally, port authorities relate to authorities that rule a set of terminals in a particular port.

Subjects were selected in North America, Europe, South America, the Middle East, and the Far East to bring complementary opinions. They also come from different backgrounds and have many years of experience in the container industry or sustainability.

3.2. INTERVIEWS ANALYSIS

During the interviews, a large diversity of factors was identified as contributing or slowing down the investment in decarbonisation technologies as anticipated in the Multi-Level Perspective on Socio-Technical Transitions theory. The author decided to use PESTLE for the first-level categorization of such factors. The PESTLE Analysis, created by Francis Aguilar, is a strategic tool to help businesses understand political, economic, social, technological, legal, and environmental impacts. It enables the industry to identify all influencing elements, break free from assumptions and adapt to reality.

The factors were also ranked and analysed based on the number of mentioned interviews. Since no list of criteria was provided, the interviewees' answers reflect only the drivers (or barriers) present in their own personal experience and professional lives.

4. RESULTS

4.1. DEMOGRAPHIC CHARACTERISTICS OF SUBJECTS

Up to 30 interviews were planned, and a total of 27 interviews were completed, aligned with the plan. The profile of the respondents is summarized in Appendix B. Most respondents are currently based in Europe (n=19) due to the concentration of equipment manufacturers and sub-suppliers and the author's access to the professionals.

Concerning the years of experience in the container industry, 74% of the respondents have more than 10 years. This indicates that the respondents were informed about the industry and had professional expertise in the field, and hence could provide credible data for this study.

Respondents were asked to auto-classify themselves in experts, influencers, or decision-makers in decarbonisation investments. They could choose more than one classification or none. Figure 5 provides the summary of their answers extracted from Appendix B. There is a fair distribution in their role in the decision-making process, with a predominance of influencers pulled by the responses of Sales/Marketing and Sustainability professionals representing 2/3.

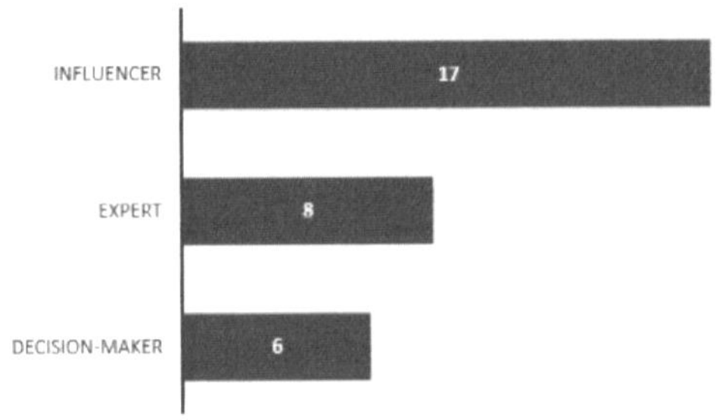

Figure 5: Auto-classification of subjects based on their role in the decision of investments.

Figure 6 summarizes the backgrounds of the interviewees. Diversity in the backgrounds, industry role, decision-making process, and geographical location also produce diversity in perspectives, making the analysis complete and more representative.

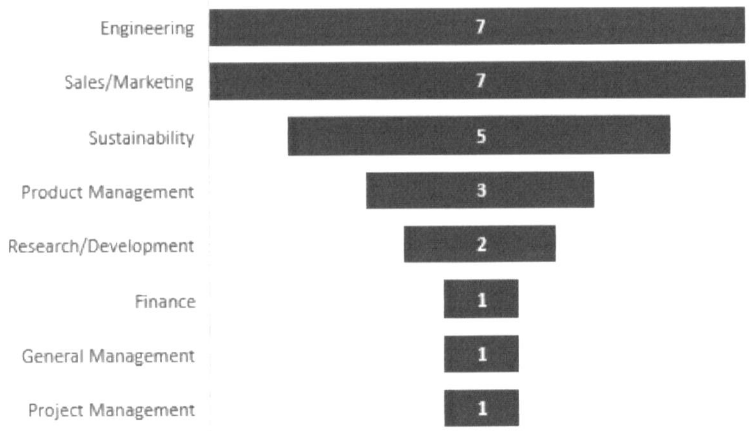

Figure 6: Backgrounds of subjects.

4.2. YEARS IN CONTACT WITH DECARBONISATION

A total of 12 subjects (44%) answered that they have been working with or discussing sustainability initiatives in their professional lives for more than 5 years. However, 19 (70%) subjects said the topic became "hotter" in the past 3-5 years, especially in decarbonising initiatives. Seventeen interviewees (63%) stated their companies had written sustainability strategies, and ten have clear goals towards net-zero emissions. Four respondents (15%) affirmed they are currently working on defining their company strategies. Two professionals confirmed their companies use external audits and/or consultancy to validate their carbon footprint calculations and the feasibility of their road maps towards net zero. Two answers also highlighted the adoption of the Science-Based Targets, an initiative to provide companies a clearly defined pathway to reduce emissions in line with the Paris Agreement goals.
Based on this evidence, we conclude that some companies may intrinsically pursue decarbonisation, but the presence of legally binding external forces led by governments, such as the Paris Agreement, is the real game-changer to drive container handling equipment manufacturers and customers to invest in solutions to reduce GHG emissions.

4.3. DRIVERS FOR AND BARRIERS TO INVESTING IN DECARBONISING TECHNOLOGIES

4.3 (a) PESTLE Analysis

The first step was to create a PESTLE Analysis to understand the drivers and barriers to investing in decarbonising technologies. It was used to cluster the factors comprehensively and add other elements mentioned during the interviews. Together, they provide a complete picture of the macro-scenario of the low-carbon transition for container terminals.

With this initial analysis of the information gathered with the interviewees, it is now possible to tackle the diversity of elements playing a role in this transition, including in political, environmental, legal, and social aspects, more than the only economic and technological approach behind the initial assumption of this study.

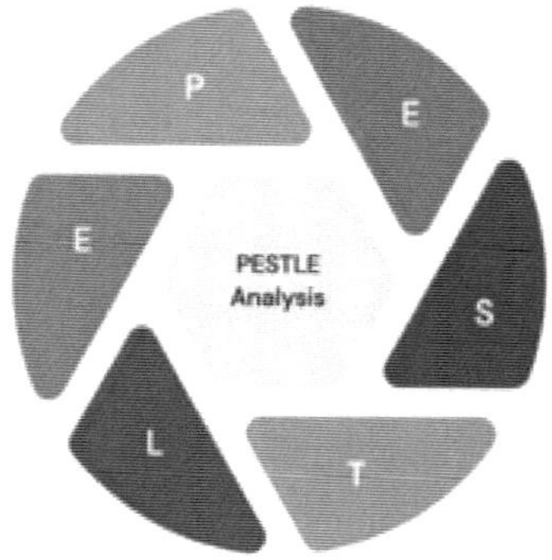

Political

- Paris Agreement commitment
- Carbon tax
- Subsidies
- More action in developed economies (e.g. Europe, USA) than in the rest of the world
- Lobby from traditional industries slowing down the change

Economic

- Green Premium still high, but decrease in price sensitivity
- Perspective of long-term returns
- Financing cost increasing for "dirty" projects
- Infrastructure cost
- Energy price
- Cash available: record profits of the industry during the Covid19 Pandemic
- Industry stability
- Availability of renewable energy and alternative fuels close to customer
- Differentiation and/or qualification factor
- Remain competitive - do or die
- New business models
- Return on investment
- Internal carbon pricing
- Competition for resources (batteries, biofuels, etc) with other industries (auto)

Social

- Green marketing/branding
- Attraction of talents
- Increasing society pressure especially among younger generation
- Media hype
- Broader internet access
- Pioneering leadership styles
- Acceptance
- Presence of sustainability in the local culture
- Educational level
- Psychological barriers: fear of the unknown, long-term planning, defensive mindset
- Change in consumption behaviour

Technological

- Electrification and hybrid solutions
- Alternative fuels (hydrogen, biodiesel, renewable diesel, etc)
- Technologies to monitor, control and reduce consumption
- Increasing investments in R&D
- Unclear direction which direction market will go
- Technological evolution/maturity
- Menu of zero emission technologies: different solutions for different markets/locations
- Shore power for vessels
- Automation
- Intrinsic limitation Heavy-lift industry
- Operational challenges (layout, modus operandi, etc)

Legal

- Local regulations limiting allowable GHG emissions
- Additional licenses and commitments for new projects being required
- Unpredictability of new regulations/requirements

Environmental

- Climate change due to global warming
- Increasing demand for electric, hybrid and zero emissions solutions
- Lack of consolidated information at industry level (container terminals)
- Public (population) health
- Moving the needle: road maps to maximize results
- Limited reserves of fossil fuels
- Other sustainability initiatives (waste, water, social, etc)
- Carbon inventory
- Certifications (Science-based Targets), external audit, expansion of scopes (1,2 and 3)

Figure 6: PESTLE Analysis. Elaborated by the author. Style adapted from Hislide.

4.3 (b) Overall Mentioning of the Factors

Rankings were built to identify the most relevant factors or tendencies mentioned by the specialists (n=27).
In the **political area**, 10 subjects brought up the geographical location poses as a decisive influencing factor impacting regulations and decarbonisation speed. They mentioned more action in the United States and Europe, although the Far East (China, Singapore) is emerging in the net-zero run.
Subsidies (n=9) made available by governments and the additional taxation on fossil fuels (n=7) were mentioned as drivers of adopting the new technologies. Even small subsidies were seen as a good sign.

> *"If energy cost is important to your business, you better prepare yourself. When the taxes have been applied, everybody will be rushing for wind turbines, solar panels, and there will not be enough. [To anticipate investments in carbon-neutral solutions] is a reasonable approach to manage your risk exposure. And the blockers are not stronger than these drivers." (Subject 27)*

> *"Today, we still can influence the politicians and the public opinion. But if we wait when the legislation comes, then we will be forced to apply some solutions we don't like." (Subject 12)*

Four subjects recognized the simple existence of the Paris Agreement with clear goals as a shifting engine.
One professional sees the lobby of the big companies, generally oil and gas, as a deceleration factor.

> *"When you are a big conglomerate always doing the same thing for the past so many years, you don't want to change... they want to make as much money as they can. They don't want to take any risk, which is understandable. They have their quarterly figures to produce." (Subject 21)*

In the **economic landscape**, as expected, the necessity of having a return on investment is a critical item noted by 60% of the respondents. However, 52% of the experts highlighted a change in the customers' price sensitivity in the past 1-2 years. Manufacturers still feel that the customer's willingness to pay for zero-emissions technologies is unclear. The concept of Green Premium was introduced by 9 interviewees, mainly from container terminals, who defended it as still a too big barrier to investments.

> *"The green premium is still too high. We still run in detaxed diesel that is cheaper than soda." (Subject 5)*

> *"We are getting closer to the breakeven, and the coin will flip in the incoming years." (Subject 23)*

Infrastructure was appointed as a big challenge (n=13) because of the cost to implement it in case of electrification of the CHE. Additionally, the availability of the infrastructure is not uniform; in some locations, there is no reliable electrical grid available or pipelines for future fuels such as green hydrogen.

> *"Infrastructure is the biggest challenge. It makes no sense to release a technology that nobody can use or wants to use." (Subject 9)*

The energy price (n=13) impacts were recognized during the interviews. Regions, where the cost of fossil fuels are too low, have fewer incentives to invest in decarbonising solutions. Moreover, the reduction in the price of renewable energy compared to 10-15 years ago was mentioned as a genuine reason to transition.
Not the less, there is still a lot of uncertainty regarding regulations and technological development, subjects defended the early adoption of the technologies due to the perspective of future returns (n=11) in the long-term (more than 10 years). On the other hand, the necessity of producing quarterly results (n=10) and measuring the success of companies are barriers.

> *"We still look at this from the financial perspective, what is the ROI, and I believe this is wrong. This is the capitalism of the '70-the '80s. We need a new picture." (Subject 12)*

Investing in the low-carbon transition was cited as a strategic advantage to differentiate from the competition (n=9) or even as a necessity to adapt to protect market share or even the entire existence of the business. Seven professionals reported that carbon neutrality should be seen as an opportunity to create new business models.

> *"If you don't do it, somebody will, and then, you have to do it anyway." (Subject 21)*

> *"For most of the companies, sustainability is a burden. For us, it is the reason of our existence. We were created from sustainability." (Subject 19)*

In recent years, with the prominent asset managers in the world declaring investments only in sustainable business, the interviewees noted a shift in the financing system (n=5). Banks ask for sustainability actions before lending the capital and, fewer and fewer institutions are willing to finance polluting projects. This leads to scarcity of capital for those projects, increasing the financing costs.
The other two motivating elements _ cash availability (n=5) and industry stability (n=1) _ appeared during the interviews. Companies with higher margins are seen as more suitable to invest in new solutions, while unstable industries create risk-averse management.

"We are in an industry that has money, stability; where technology is not that difficult to get to. So, if we are not moving, we are idiots." (Subject 15)

The establishment of internal carbon pricing (n=4) is seen as a motivation aspect inside corporations towards decarbonisation in future investments.

Conclusively, two experts recognized the CHE industry is small and is struggling with the competition with the giant auto industry (cars and trucks) for resources like batteries and biofuels.

In the **social field**, the leadership style, and the organizational culture (n=16) present in companies in sync with the personal beliefs of the professionals (n=16) and awareness of global warming are, by far, the drivers for change.

"Organizational culture is important to move the sustainability topic forward. Where in some companies nobody talks about it or the Sustainability Manager is on the corner, in others, it is in everybody's minds, in every meeting." (Subject 2)

Surprisingly, the pressure coming from society (n=12) scored high as a motivator for investments showing the power of public opinion on the low-carbon transition. Other motivators were green marketing (n=10), the local culture (n=7), change in consumer behaviour (n=2), and internet outreach (n=2). Companies want to be responsible and implement what they market and build a reputation out of that. Broad internet access allows the entire population to have access to accidents and pollution events in real-time, something that did not happen 10-15 years ago, and leaves companies exposed.

"We drink our own champagne." (Subject 13)

"We put the money where our mouth is." (Subject 23)

Psychological barriers appeared in 8 interviews in the form of "fear of the unknown," "natural human difficulty in long-term planning," and "defensive mindsets."

"Everything that seems to cannibalize existing business will encounter in large organizations some enthusiasts and lots of people that are pushing it back. It is just defensive. It has nothing to do with strategy; it is how people react." (Subject 27)

Singularly, subjects said that having kids is one of their main reasons why sustainability is a focus in their professional lives or proposed the exact opposite, the lack of concern with future generations (n=7). And education (n=3) was listed as part of the pressure that the new generations can exercise approaching a carbon-neutral society.

"Nobody is concerned with the next generations, how to make a better world together. It is all about money." (Subject 14)

Touching the **technological aspects**, electrification of equipment and hybrid solutions (n=20), alternative fuels (n=15), renewable energy (n=10), and shore power (n=4) are the leading technologies off the top of the heads of the specialists. Electrification of cranes is a mature and well-known technology but requires considerable investments in infrastructure and the availability of renewable energy to compose the low-carbon transition. Hybrid solutions are newer and have become popular in the past 5 years with the decrease in battery prices. Shore power, provision of shoreside electrical power to a ship at berth, is reported as a critical investment if container terminals want to reduce their carbon footprint. Unexpectedly, efficiency technologies to optimize electricity and fuel consumption at low costs and potentially fast implementation, _ "quick wins", _ were cited in only 4 interviews.

Seventeen interviewees acknowledged that many technologies need more maturity before being chosen during the planning of greenfield projects or being adopted on a big scale by customers. One person confirmed the increase in research & development investments, but there is a natural limitation of resources (staffing, budget, and time).

"If you start now, you are already years late." (Subject 4)

There was the acknowledgment that it is unclear the direction the market and the technologies will go (n=5) and, most likely, a combination of solutions (n=5) will be available for different markets.

"There will not be a single solution; there will be a combination of solutions since not all commodities are available everywhere. The main obstacle is to find the right product at the right time in the right market." (Subject 9)

Other challenges to adopting the new low-carbon technologies are operational challenges (n=3), primarily related to the charging times of horizontal rolling equipment and intrinsic limitations of a heavy-lifting industry (n=1) that faces developing and maturing alternatives. Automation (n=3) was cited as the speeding-up agent of the electrification and hybrid solutions mainly due to system communication requirements.

The existence of regulations limiting GHG emissions (n=17) and the additional requirements that can be imposed in developing projects (n=5) represent relevant drivers for companies to start investing in alternative solutions to fossil fuels. The unpredictability of the new regulations that will come up (n=7) is considered a relevant business risk or opportunity depending on the professional view.

"Nobody takes it seriously until the imposition comes." (Subject 12)

"Politicians are making decisions without thinking it through the end." (Subject 6)

For the **environmental factors**, it was frequently mentioned (n=15) the necessity of having a road map to orientate the organizations to succeed in the low-carbon transition. To comprehend what "moves the needle" and prioritize the projects to maximize the environmental return with the available resources. Professionals mentioned the expansion of the emission reduction scopes (n=2) from only internal operational emissions (scope 1) to cover the entire value chain _ suppliers and customers emissions (scopes 2 and 3). They also referred to the utilization of external consultancies (n=2) and the adoption of the Science-Based Targets (n=2) as actions to help overcome the barriers to becoming carbon neutral. One sustainability professional highlighted that having a carbon inventory is the first step to start the change.

"Some ports may not even have an emissions inventory to tell them where to prioritize their time. Where are the greatest emissions coming from? And you would want to do that. [...] A comprehensive master plan showing which pieces of zero-emissions equipment the terminals are going to invest in, what are their best options, what is the most cost-effective way to get to their final goal, what infrastructure needs to go in, where do we need upgrades. [...] When these plans are finalized, the terminal is best positioned to go after funding." (Subject 22)

Some subjects affirmed that climate change itself (n=4) is a sufficient reason to invest in decarbonisation. The limited reserves of fossil fuel and the certainty that we need to find alternative energy sources were cited by 3 people.

The increase of demand for zero-emission solutions (n=10) is a significant driver for manufacturers and sub-suppliers to invest in developing new products. But container terminals complained about the lack of consolidated information (n=5) to guide them on what is happening at the industry level and the alternatives.

"If you don't have the knowledge and know the technologies, you may miss an opportunity. You don't have good channels of information. You get pieces of information from suppliers, media, and you try to connect the dots and judge where we might be in the future." (Subject 14)

Two subjects indicated the health problems of the population living around terminals forced them to abandon the utilization of fossil fuels. Finally, the other 2 mentioned that if you are already working on other sustainability goals, decarbonisation naturally will be on the agenda.

"There are real negative impacts in the community, and those are well documented. The most critical is the increased risk of cancer for people living around the port. The driver is very real. They are a disadvantaged community. They have lower socioeconomic status, predominately people of colour, who aren't reaping the benefits of the port and are suffering the negative impacts. Significant higher levels of asthma. It is not just an organization that wants to have ambitious goals. It is about reputation, but the driver is public health." (Subject 22)

Additionally, to the PESTLE categories, some **internal factors** of the organizations were mentioned. The most relevant one was the strategic decision of embracing the net-zero goals and rolling out the strategy to the entire organization (n=10). Correlated is the shareholder and/or the headquarter role (n=7) in coordinating the implementation in big groups.

"Strategy is implemented by allocation of capital. If you want to protect something that exists and develop something new at the same time, it is extremely difficult for any organization. You cannot adapt fast enough." (Subject 27)

Lastly, increasing collaboration in the industry (n=7) is a must-do to speed up the evolution of the technology and its adoption.

"A lot of the clients also have ships and more bargain power. Clients should collaborate more with port equipment manufacturers; the client's own experience and knowledge can help the development of the products and technologies. What are the actions the clients are taking? Are there workshops?" (Subject 1)

"Few companies (6-9) controls 80% of the global transportation. I believe they have an obligation to cooperate with governments. It is very easy to establish universal standards; it is a golden opportunity. I miss that terminal operators, and equipment manufacturers are not proactive in looking for joint ventures to look for solutions now. We should not wait until 2030. Today, it is total chaos. Everybody is looking at the other one. Nobody is acting." (Subject 12)

"What would our portfolio look like in the next 2-4 years? What do customers want? I don't know, but the customers don't know either." (Subject 23)

Figure 6 shows the overall top 12 factors in the decarbonisation process of container handling equipment.

TOP 12		
PESTLE	**Factor**	**TOTAL**
Technological	Electrification/hybrid solutions	20
Legal	Local regulations limiting allowable GHG emissions	17
Technological	Technology evolution/maturity	17
Economic	Economic viability/ROI	16
Social	Leadership/organizational culture	16
Social	Awareness/Personal Belief/Character	16
Environmental	Moving the needle/Internal Project Prioritization/Road Map	15
Technological	Alternative fuels	15
Economic	Price sensitivity/customer willingness to pay	14
Economic	Infrastructure cost & availability	13
Economic	Energy price (fossil fuel, electricity, etc)	13
Social	Pressure from society/social responsibility	12

Figure 7: Top 12 mentioned factors influencing investments in low-carbon technologies.

Figure 7 summarizes the model formulation and implementation perspective of the criteria identification towards the low-carbon transition used for this study.

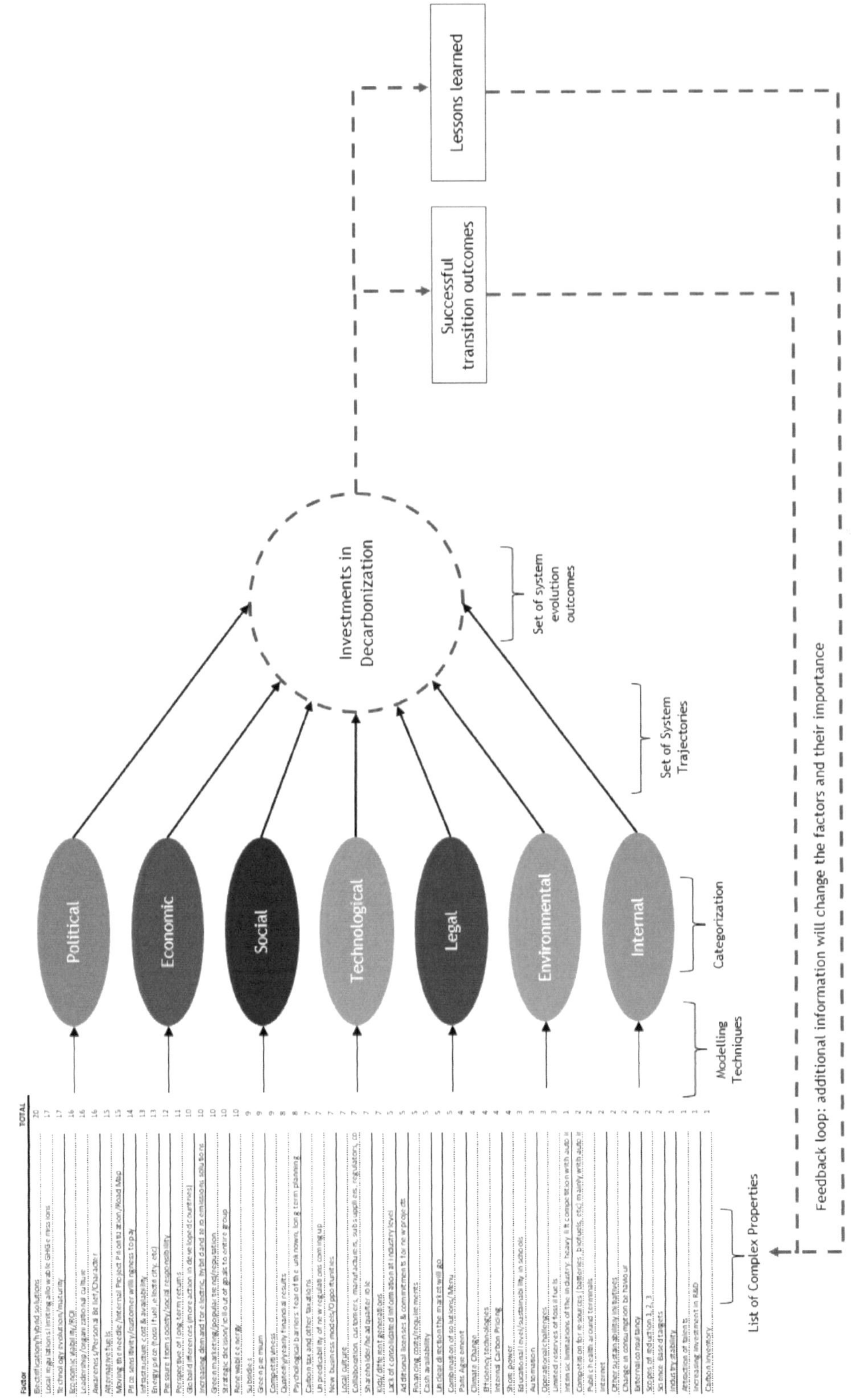

Figure 8: Drivers for investing in the low-carbon transition: model formulation and implementation process. Adapted from [24].

4.4. LEADERSHIP STYLES AMONG THE FRONTRUNNERS

The leadership and the organizational culture were ranked in the top 5 as drivers to invest in solutions that reduce emissions. This phenomenon appears in the literature, and it was the reason why the author dedicated one question to investigate what is behind frontrunner leaders.

During the interviews, 4 leadership styles were identified behind organizations that have an appetite for the new and tend to be seen as successful frontrunners in the industry: visionary, adaptative, conscious, and innovative.

Visionaries are leaders who can identify tendencies or future opportunities not spotted yet by others. They focus on the long-term returns and believe the investment will pay off.

> *"I know a chairman that says: 'This will be a good business'. He is anticipating a competitive advantage. He can be a visionary, but he can also have privileged information (in a good way)." (Subject 13)*

Adaptative leaders respond promptly to a necessity; it can be due to public health impact, new regulations, customer demand, pressure from lenders, or just because of the headlines. They read the external signs and decide to act fast.

> *"We have the demand from big clients. We need to comply immediately if we want to keep market share." (Subject 1)*

> *"[...] we invested a lot, and we succeed. That is it. It was a necessity for us. Otherwise, we would die in 20 years." (Subject 21)*

It was also possible to identify leaders that were genuinely concerned about the environment, climate change, and the impact on future generations. They align personal beliefs with their professional life as an opportunity to make a more significant impact.

> *"I genuinely believe there are leaders out there that there are really concerned with the environment [...]. Not all of them. [...] I hope they are the majority." (Subject 6)*

> *"Some of these leaders are more conscious about environmental impact, and they are in the position of doing their part. I believe it is coming from their background, their education, their personal beliefs, their personal life, and they want to reflect that in their professional life." (Subject 7)*

Further, innovative leaders are risk-takers by nature. They like to experiment and want to be seen as creatives and originals. They build their competitive advantage by being ahead of competitors with new products and solutions. Sustainability is only a part of that approach.

> *"The people who are motivated to explore the newer technologies and willing to take the technological risk over the business risk, they are innovators [...]. The selection of horizontal transport with battery took a lot of courage and due diligence to understand the solution. The diesel is the convenient and lazy way, safe; it has been there for decades. The electric one has a technological risk [...]. Those kinds of things need some courage of the people making the decision, not only not to be risk-averse." (Subject 8)*

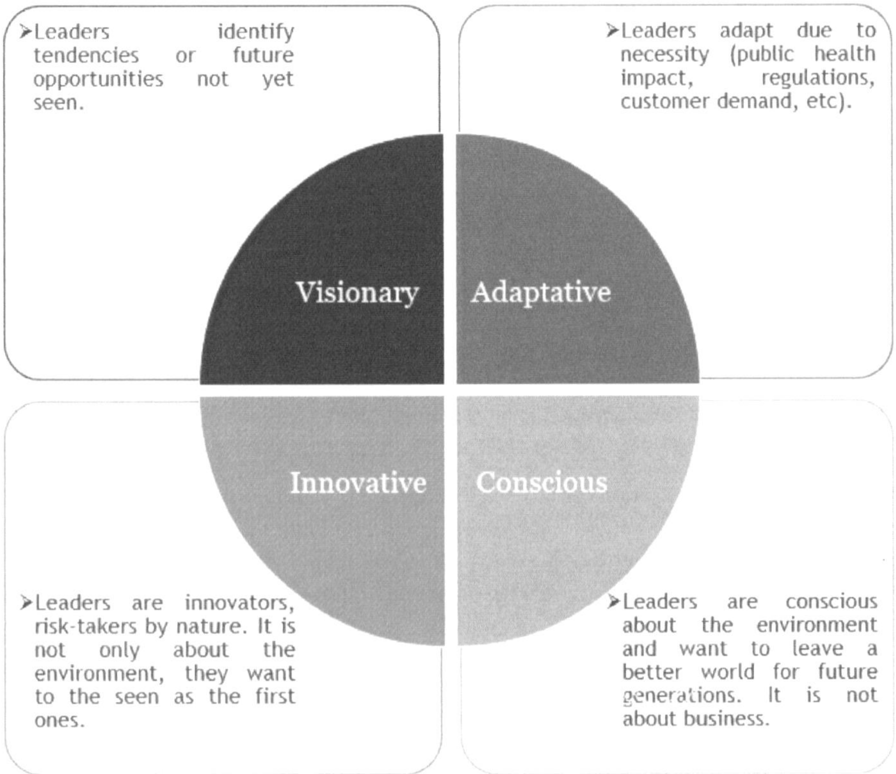

Figure 9: Leadership styles identified in frontrunners during the interviews. Elaborated by the author.

5. CONCLUSIONS

The concern around supply chains to become carbon neutral intensified in the past years since the Paris Agreement. Container terminals are a vital link with a relevant role in shipping decarbonisation. The decision-making environment is complex and fraught with uncertainty. Information is scattered, and good decision support is required.

There are several obstacles to overcome when implementing technologies to minimize emissions in container handling equipment. The criteria differ, making comparison difficult and, more significantly, leading to findings that are in contradiction. This paper investigates the key drivers influencing the decision-making in a multi-stakeholder participatory approach in the container industry.

The interviews revealed the importance of various socio-technical structures in guiding the system's progress down diverse paths. Internal and external influences have shaped diverse players (ideologies and purposes), technologies, and regulations within these industry configurations, triggering various types of transitions and bringing the system to its current state.

This study concludes that time is short; we should act now. Actors cannot adopt a passive posture and wait for technology to mature. The container industry is profitable and stable, but there is still an innovation gap that the interviewed experts believe can be overcome with coordination. Coordination among container terminals, equipment manufacturers, components sub-suppliers, regulators, and energy providers is vital to developing more robust decarbonising solutions in a shorter time.

The concept of economic viability is changing, or at least losing importance, with other metrics, such as carbon footprint, starting to play a relevant part in procurement. There is a perception that the capitalism of the '70-'80s is over and giving way to stakeholder capitalism; in other words, the reason to exist of a company is not only profit anymore, but to improve society and protect the environment.

Other elements besides technology maturity and return on investment were also meaningful. The public opinion and the media exercise massive power over companies and politicians. They gear the way companies attract talents, engage their employees, and help build new leadership styles and organizational cultures. The way people vote, the pressure from non-governmental organizations, and governments drive the advancement of new legislation, taxation, and the availability of subsidies.

Parents are concerned about the planet they will leave to their kids and are constantly influenced by them. The Greta Thunberg Generation is not only our future CEOs and parliament members; every million students, there are six million adults behind them; their influence is massive.

The infrastructure availability and respective cost will pave the way for master plans and road maps in organizations committed to the low-carbon transition. Different zero-emissions technologies will be implemented in different parts of the world, respecting local characteristics. The institutions that can have written strategies, clear goals, master plans, road maps audited by external specialists that can roll it out to all business units will have more chances of successfully positioning themselves as frontrunners.

Ultimately, it all comes down to people. In essence, leadership is a key interpreter of how the wider complex adaptive systems environment interacts internally with the organization. Pioneer leaders are motivated by a sharp vision of the future, a genuine environmental conscience, a survival instinct, or an innovative natural acumen behind all these organizations. They will have to use the ability to navigate through the complexity of sustainability, the complexity of leadership, and factors of complex problem solving that we are still endeavouring to describe.

It is estimated that some 80 percent of all goods are carried by sea. Global maritime container trade is estimated to account for around 60 percent of all seaborne trade in terms of value, which was valued at approximately 14 trillion U.S. dollars in 2019 [37]. The container handling equipment is the biggest energy consumer (electricity, diesel, gas, etc.) in container terminals (excluding potential vessels connected in shore power and the occupancy of reefer plugs) and, therefore, a relevant topic in the decarbonisation of supply chains. The low-carbon transition in the sector will not be straightforward; however, it is possible to evolve if opportunities can be identified along with the socio-technical analysis that supports them.

"We don't have all the answers today. There is still this innovation gap. But it does not impact our ambition." (Subject 26)

"Managers must understand we will have to make sacrifices in the future. Our life will change, whether you want it or not." (Subject 12)

6. ACKNOWLEDGEMENTS

I want to express my gratitude to ESMT Berlin for its support in making this article a reality. This work would not have been possible without my interviewees' sincerity, availability, and contributions. Finally, my beloved husband deserves all the glory for always motivating me.

7. REFERENCES

1. 'Can the global shipping industry chart a path to decarbonisation?', *Oxford Business Group 20,* 2021.

2. WHIELDON, E., 'Your climate change goals may have a maritime shipping problem', *S&P Global,* 2021.

3. MERK, O., 'Shipping Emissions in Ports', *International Transport Forum Discussion Papers,* 2014.

4. SPENGLER, T., & WILMSMEIER, G., 'Sustainable Performance and Benchmarking in Container Terminals _ The Energy Dimension', *Green Port: Inland and Seaside Sustainable Transportation Strategies,* 2019.

5. 'The Paris Agreement', *United Nations Climate Change web site.*

6. FINK, L., 'A Fundamental Reshaping of Finance', *BlackRock web site*, 2020.

7. MAINWARING, S., 'Purpose At Work: Why BlackRock's CEO's Letter Is A Wake-Up Call To All Brands', *Forbes*, 2020.

8. 'Cargo-Handling Equipment', *Clean Air Action Plan web site.*

9. FUNG, F., et al, 'Prevention and Control of Shipping and Port Air Emissions in China', *NRDC - Natural Resources Defense Council,* 2014.

10. 'Ports of Auckland – Zero Emissions 2040', *Sustainable World Ports web site.*

11. 'Energy Transition - Strategy and Vision', *Port of Rotterdam web site.*

12. 'Objetivo Puertos Cero Emisiones'. *Dia del Medio Ambiente web site.*

13. 'Climate Action', *Port of Antwerp web site.*

14. SAUL, J., 'UK's Peel Ports pushes to be net zero of carbon emissions by 2040', *Reuters*, 2021.

15. SCHULTZ, F., 'German cabinet agrees CO2 price of €25 from January 2021', *Euroactiv*, 2020.

16. RODRIGUE, J.-P., 'The Geography of Transport Systems', *Routledge*, 2020.

17. UNCTAD, 'Review of Maritime Transport 2021', *United Nations Conference on Trade and Development*, 2021.

18. OECD, 'Decarbonising Maritime Transport. Pathways to zero-carbon shipping by 2035', *International Transport Forum 86*, 2018.

19. TANER, M. et al, 'Layout analysis affecting strategic decisions in artificial container terminals', *Computers & Industrial Engineering*, 2014.

20. GEELS, F. W. et al, 'The Socio-Technical Dynamics of Low-Carbon Transitions', *Joule*, 2017.

21. JANO-TIO, M. A., & CRAWFORD-BROWN, D., 'Socio-technical analysis of the electricity sector of Mexico: Its historical evolution and implications for a transition towards low-carbon development', *Renewable and Sustainable Energy Reviews 55*, 2017.

22. NICHI, J., & JUNIOR, J. Z., 'Governança multinível: o papel da sociedade civil na transição para um transporte de baixo carbono em sistemas sociotécnicos', *Brazilian Journal of Development*, 2021.

23. SCRASE, I., & SMITH, A., 'The (non-)politics of managing low carbon socio-technical', *Environmental Politics*, 2009.

24. PETER, C., & SWILLING, M., 'Linking complexity and sustainability theories: Implications for modeling sustainability transitions', *Sustainability*, 2014.

25. ASHRAFI, M. et al, 'Toward a harmonization of sustainability criteria for alternative marine fuels', *Maritime Transport Research 3*, 2022.

26. CHRISTODOULOU, A., & CULLINANE, K., 'Potential for, and drivers of, private voluntary initiatives for the decarbonisation of short sea: shipping: evidence from a Swedish ferry line', *Maritime Economics and Logistics*, 2020.

27. LU, C.-S. et al, 'Identifying crucial sustainability assessment criteria for container seaports', *Maritime Business Review*, 2016.

28. LIRN, T. C. et al, 'Green performance criteria for sustainable ports in Asia', *International Journal of Physical Distribution and Logistics Management*, 2013.

29. MARCHET, G. et al, 'Environmental sustainability in logistics and freight transportation: A literature review and research agenda', *Journal of Manufacturing Technology Management*, 2014.

30. 'Do sustainable products require sustainable leaders? How managers can steer a course between people, profit and the planet', *Development and Learning in Organizations*, 2011.

31. KNIGHT, B., & PATERSON, F., 'Behavioural competencies of sustainability leaders: an empirical investigation', *Journal of Organizational Change Management*, 2018.

32. HANEY, A. B. et al, 'Making it personal: developing sustainability leaders in business', *Organization and Environment*, 2020.

33. MARATHE, G. M. et al,'Is management education preparing future leaders for sustainable business? Opening minds but not hearts', *International Journal of Sustainability in Higher Education*, 2020.

34. HIND, P. et al, 'Developing leaders for sustainable business', *Corporate Governance: The international journal of business in society*, 2009.

35. THAKHATHI, A. et al, 'Sustainability Leaders' Influences Strategies for Institutionalising Organizational Change towards Corporate Sustainability: A Strategy-as-Practice Perspective', *Journal of Change Management*, 2019.

36. GROVER, P. et al, 'Influence of political leaders on sustainable development goals - insights from Twitter', *Journal of Enterprise Information Management*, 2021.

37. PLACEK, M., 'Container shipping - statistics & facts', *Statista*, 2021.

8. AUTHORS BIOGRAPHY

Carolina Parro has 11 years of experience in container logistics and ports mainly in operational and technical areas. She holds an Executive MBA with honours from ESMT Berlin, Germany, and a Production Engineering bachelor's degree from Federal University of Sao Carlos, Brazil.

APPENDICES

APPENDIX A: SURVEY QUESTIONARY

1- How long have you been in the container handling equipment industry?

2- How long have you been working with/discussing ESG initiatives?

3- Does your company have a written strategy with clear goals to reduce emissions? If not, why not? If yes, when was it released?

4- So far, do you believe your company takes the necessary actions to reach the goals? If not, why not?

5- Does your company invest in new technologies to reduce emissions in its products or the supply chain?

6- In your experience, what are the barriers to your company's investment in new technologies and products? And, on the other hand, what are the motivators?

7- In your experience, what are the barriers to your CLIENTS investing in new technologies and products? And, on the other hand, what are the motivators?*

8- Why are some leaders pioneers (and consequently their companies) in adopting technologies that reduce emissions?

9- Do you consider yourself an expert, an influencer, or a decision-maker when investing in technologies and products that can help reduce emissions?

10- Is there anything worth mentioning that I did not ask? Final comments?

* Question number 7 was only applied in subjects that were categorized as "Equipment Manufacturer", "Sub-supplier" or "Utility Supplier".

APPENDIX B: PROFILE OF THE SUBJECTS

	Geographical Location	Category	Years in the Industry	Dept.	Auto-classification*
Subject 1	Far East	Equipment Manufacturer	18	Sales/Marketing	Influencer
Subject 2	Europe	Container Terminal	18	Engineering	Influencer
Subject 3	South America	Container Terminal	32	Engineering	Expert
Subject 4	Europe	Equipment Manufacturer	32	Sales/Marketing	Influencer
Subject 5	Europe	Container Terminal	3	Engineering	Expert/Influencer
Subject 6	Middle East	Container Terminal	30	Engineering	Expert
Subject 7	North America	Container Terminal	25	Engineering	Decision-maker
Subject 8	North America	Container Terminal	16	Engineering	Decision-maker
Subject 9	Europe	Equipment Manufacturer	10	Sales/Marketing	Influencer
Subject 10	Europe	Equipment Manufacturer	6	Product Management	Expert/Influencer
Subject 11	Europe	Container Terminal	17	Sustainability	Influencer
Subject 12	Europe	Container Terminal	25	General Management	none of the 3
Subject 13	Europe	Subsupplier	22	Sales/Marketing	Influencer
Subject 14	Europe	Container Terminal	24	Project Management	Decision-maker
Subject 15	Europe	Container Terminal	6	Finance	Influencer
Subject 16	Europe	Container Terminal	25	Engineering	Influencer
Subject 17	Europe	Equipment Manufacturer	22	Sales/Marketing	Influencer
Subject 18	Europe	Container Terminal	0	Sustainability	-
Subject 19	Europe	Subsupplier	2	Sales/Marketing	Influencer
Subject 20	Europe	Equipment Manufacturer	12	Product Management	Decision-maker
Subject 21	Middle East	Subsupplier	30	Sales/Marketing	Expert/Influencer
Subject 22	North America	Port Authority	7	Sustainability	Expert/Influencer/Decision-Maker
Subject 23	Europe	Equipment Manufacturer	31	Product Management	Decision-maker
Subject 24	Europe	Container Terminal	16	Sustainability	Influencer
Subject 25	Europe	Utilities	14	Research/Development	Expert/Influencer
Subject 26	Europe	Equipment Manufacturer	18	Sustainability	Influencer
Subject 27	Middle East	Utilities	25	Research/Development	Expert

* Auto-classification when interviewees were asked if they considered themselves experts, influencers, or decision-makers when investing in technologies to reduce emissions.

THE SILK ALLIANCE: EXPERIENCE AND INITAIAL LESSONS FROM A GREEN SHIPPING CORRIDOR CLUSTER

C Raucci, A Karan, and S Balani, Lloyd's Register Maritime Decarbonisation Hub, UK

SUMMARY

This paper briefly introduces the green shipping corridor cluster, the Silk alliance, its evolving scope, and its overall approach. It describes the progress made so far, and the initial lessons learned during the first phase of this initiative. It concludes with some key remarks from our experience as a facilitator and from the role we have as a decarbonisation hub.

1. INTRODUCTION

The shipping industry is in a crucial historical moment where another energy transition is going to shape the industry as we know it today. The drivers and challenges are different from previous energy transitions; however, the long-term system change may happen similarly as it has been observed in the past; starting from niche markets until their expansion merges and substitute with the current regime. There is still uncertainty around the shipping energy transition pathway due to future technology developments, policy, and macro trends that could influence the way the maritime industry could evolve. Moreover, upcoming years could see important changes at IMO with the revision of the GHG strategy potentially aiming for full decarbonisation by 2050, and key conversation around the inclusion of fuel standards, LCA, and carbon tax. Inevitably all of this will give another important signal to the industry.

Despite these uncertainties, there is an increasing urgency in acting now and mitigating climate risks. The focus is on an actionable plan that can drive the implementation of decarbonisation solutions; therefore, the attention is going on 'First movers'. Stemming from the Clydebank Declaration at COP26 in November 2021[1], Green corridors are now most commonly defined as "a shipping route on which the technological, economic, and regulatory feasibility of the operation of zero emissions ships is catalysed through public and private actions" [2]. Green corridors can therefore be seen as niche markets where first movers can create protected spaces that allow the experimentation with the co-evolution of technology, user practices, and regulatory structures for zero emissions ships. The United States and Norway organized the Green Shipping Challenge for COP27 where countries, ports, and companies made more than 40 major announcements on issues such as innovations for ships, expansion in low- or zero-emission fuels, and policies to help promote the uptake of next-generation vessels [3]. Almost 50% of the announcements included the concept of 'green shipping corridor'.

The Silk alliance, established by LR Maritime Decarbonisation Hub (LR MDH), is an initiative of 12 leading cross-supply chain stakeholders to develop a fleet fuel transition strategy that can enable the establishment of a highly scalable green corridor cluster. This paper will briefly describe the underlined approach used e.g. the First mover Framework [4], and it will focus on the experience gained, and the initial lessons learned during the first phase of this initiative. We hope that by sharing this information we encourage greater transparency around green corridor initiatives which would provide direction to learning processes, attract attention, facilitate interactions, and legitimate continuing nurturing for such type of initiative.

2. AN ALLIANCE FOR A GREEN SHIPPING CORRIDOR CLUSTER

2.1 INTERPRETING THE DEFINITIONS OF GREEN SHIPPING CORRIDORS

The concept of "Green shipping corridors" has first been put forward in [2] and since then has been subject to different interpretations leading to different definitions and approaches [5, 6]. Today there are more than ten green shipping corridors initiatives that can be categorised in different ways, whether it be the focus on specific fuel/technologies or the type of ship, the number of ports involved as well as the role they should have within the energy transition. Despite the different definitions of green corridors, we think that the building block of a green corridor initiative should be:

a) participants should cover the entire supply chain involving private and public organisations,
b) their role is to spur early and rapid adoption of fuels and technologies in a way that enables the 'spillover' effect supporting the sector-wide decarbonisation

Fundamentally a green shipping corridor is a way to incentive stakeholders to come together and act as 'first movers' on a specific maritime application to start the energy transition. The most important factor is that there is a concentration of enough demand for new fuels in a relatively small number of ports so that the inland fuel supply infrastructure can be put in place to serve this demand. The terms 'corridor' and 'route-based' can often be a limiting factor because first movers can act at a regional level, as we demonstrate in the case of the Silk Alliance. We use the term "green shipping corridor cluster" to highlight the regional scope of this initiative which can involve one or several ports as long as the regional activity captures sufficient fuel demand. Given the international nature of the shipping sector, and "point-2-point" routes or a "route-based" focus are unlikely or very difficult to capture the aggregated demand and required scale to unlock investments in the necessary fuel supply and distribution infrastructure. The "green shipping corridor cluster" instead is an initiative for first movers at a regional level, covering multi-countries, and potentially multi-sectors with the shipping sector acting as a key enabler for a wide energy transition.

2.2 THE EVOLVING SCOPE OF THE SILK ALLIANCE

The Silk Alliance started as:

"a project focusing on delivering a decarbonisation plan for a fleet of small-medium containerships operating in a route between Singapore and another port"

and it has now evolved into:

"a programme focusing on creating and implementing a green corridor roadmap based around a decarbonisation plan for a fleet that predominantly operates and refuels in Singapore, which includes different vessel types trading across multiple ports from Pacific Islands, South East Asia, North Asia to East Africa"

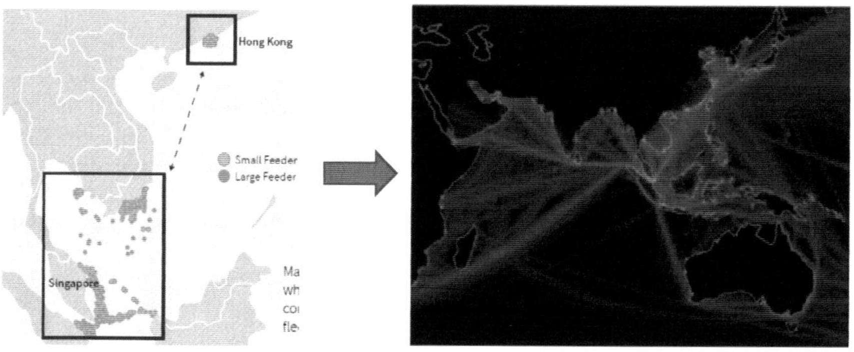

Figure 1: On the left, a representation of the initial concept of a green corridor involving feeders around Singapore as per [4]. On the right, a representation of the activities of the ships that could be part of the Silk Alliance and the region under the scope.

This change in scope means:

a) the Silk Alliance aims to focus on delivering and implementing a roadmap with a phase of continuous monitoring and refining of the roadmap
b) the Silk Alliance is not only looking at containership feeders exclusively operating around Singapore, instead, it is looking at ships with a specific common denominator: a fleet that mainly operates in the same region (regional bounds stretching to the Pacific Islands, South East Asia, North Asia to East Africa) and bunker in Singapore.
c) The Silk Alliance is looking at inland fuel infrastructure from countries with the potential to competitively produce these fuels and are within this region. For example, as found in [7], there are different countries that have the potential to produce hydrogen-based fuels in this region.

3. THE OVERALL APPROACH

The overall approach is based on the "First Movers Framework" developed in [4]. This framework aims to guide the private and public stakeholders in the supply chain to systematically evaluate decarbonisation strategies for a specific fleet. The framework is designed with the view that both sides, the fuel supply, and the fleet should work in tandem to support the energy transition along a shipping 'corridor' to ensure its success.

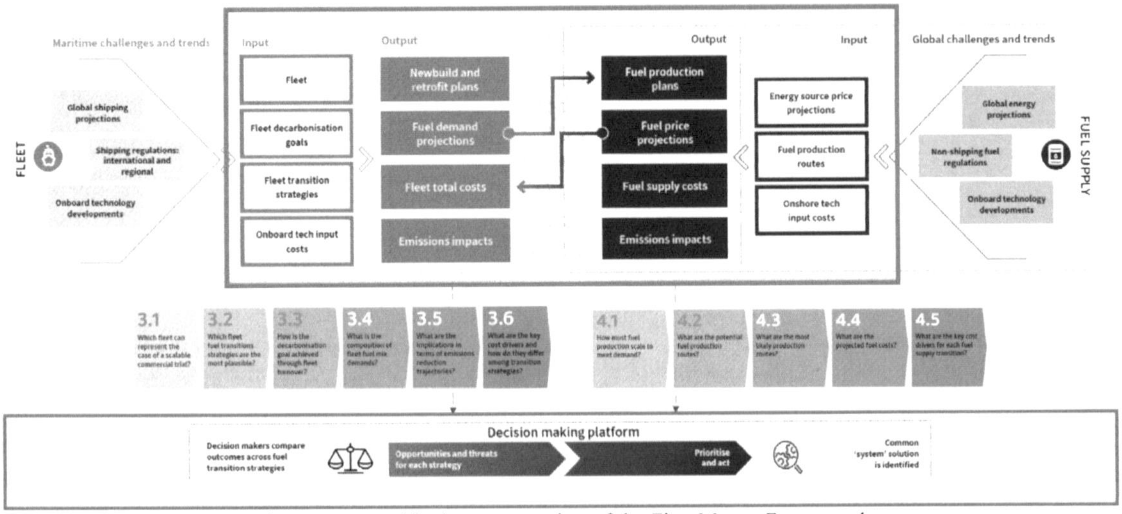

Figure 2, a graphical representation of the First Mover Framework

The framework was first developed for an explorative pre-feasibility study where a first version of different models was developed. As we are applying the framework to the Silk Alliance project, the models of this framework are being further developed and refined based on the ongoing work. The fleet transition model described in the following section is an example of further development.

The First Mover Framework can be divided into three key parts:

- System thinking stage (yellow boxes in Figure 2 and 3): brings together the experiences of all the Members; learnings from existing literature and the industry to bolster the Alliance's common understanding of the challenges. Drawing on these collective experiences the members identify the key questions to be addressed in the following stage.
- Foresight stage: First, key input and scenarios are defined, second a quantitative analysis is undertaken to provide new insights and evidence regarding the costs and benefits of different decarbonisation strategies.
- Co-Creation stage: this final stage aims to create consensus for a common strategy to build an implementation plan through collective a road mapping exercise.

We translate these parts into an agreed work process composed of six different activities as shown in figure 3.

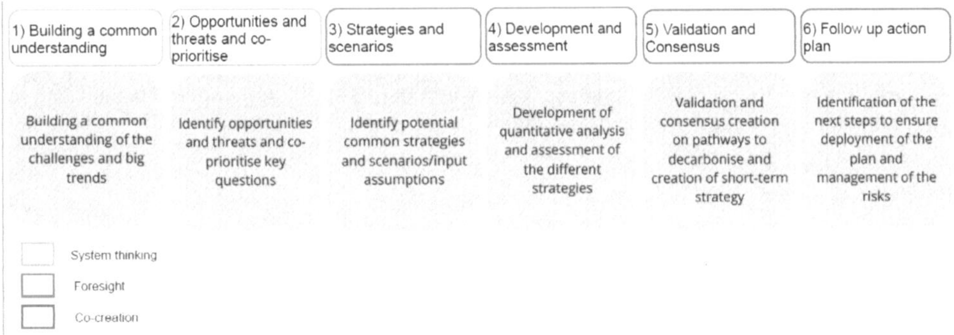

Figure 3, graphical description of the process composed of six different activities to implement the First Mover Framework

4. THE PROGRESS

4.1 BUILDING A COMMUNITY

Members of the Silk Alliance were invited to share their understanding of the challenges that have curbed investments in decarbonisation demonstration projects. These challenges were prioritised, and key questions were identified. This was done through a series of in-person workshops and working group calls. This stage allowed the creation of a new communications channel among the members that were not previously collaborating and the identification of about 75 challenges that were then translated into a series of key questions grouped into three themes:

1. Alignment with expected policy considers potential changes to the IMO level of ambition, the inclusion of well-to-wake methodology, and carbon taxes.
2. The pace of change: considers implications of different fuel uptake and fleet-turnover speeds. Fleet decarbonisation can occur at a gradual pace, with transition fuels and production methods, or it could be accelerated to skip only to the greenest fuels.
3. The fuels: considers the various fuels available, as individual choices, or as a multi-fuel transition

4.2 THE SCENARIOS-BASED ANALYSIS

Scenario-based analysis using models is a fundamental tool for climate actions. Climate change science is largely based upon models which have become indispensable for the formulation of climate policy as well business decisions. In this stage, we used models to assess and compare different fuel transition strategies. Firstly, several scenarios and input assumptions were defined. Secondly, the baseline fleet was identified, and finally, fleet turnover and costs and benefits analysis were carried out.

4.2 (a) Definition of the scenarios

Smaller working groups and a virtual workshop were held to agree on several input assumptions. These include several fixed assumptions such as emissions factors, fuels characteristics, engine, and tanks, fuel prices, and annual growth of transport demand, as well as key input variables: emissions reduction goals, carbon prices, the share of the fuels over time, the relative role of early scrap vs retrofitting as summarised in table 1.

Table 1, description of the input assumptions

Input assumption	**Type**
Emissions factors	Fixed for all scenarios
Fuel characteristics	Fixed for all scenarios
Fuel prices	Fixed for all scenarios
Engine and tanks (costs, performance)	Fixed for all scenarios
Transport demand annual growth	Fixed for all scenarios
Emissions reduction goals	2 different cases
Carbon prices	3 different cases
Share of fuels uptake over time	16 different cases
Relative role of early scrap vs retrofitting	4 different cases

To construct the scenarios, we used the process described in [8]. We identified several scenarios by combining the contrasting outcomes of each theme identified in the previous stage. Following this process, we identified 64 different scenarios.

4.2 (b) The baseline fleet

A smaller working group was formed to identify the baseline fleet under the scope and gather fleet data from the shipowners. The baseline fleet is not necessarily every vessel owned and operated by the shipowner's members, as the fleet selection must ensure stable and reliable energy demand can be inferred or derived to give fuel producers the certainty to invest in dedicated production infrastructure.

By processing the fleet data provided by the members and the historical movements of the fleet using AIS data the baseline fleet was defined as:

- vessels already predominantly operating in the identified region and already refuelling in Singapore
- vessels mainly operating in the region (up to 75% of the time in the region depending on owners) and likely to refuel in Singapore
- the identified region was chosen considering that the ships can operate their roundtrips and still have the ability to refuel in a single point

The catchment area of the Silk Alliance as well as the major concentration of activity are represented in figure 4.

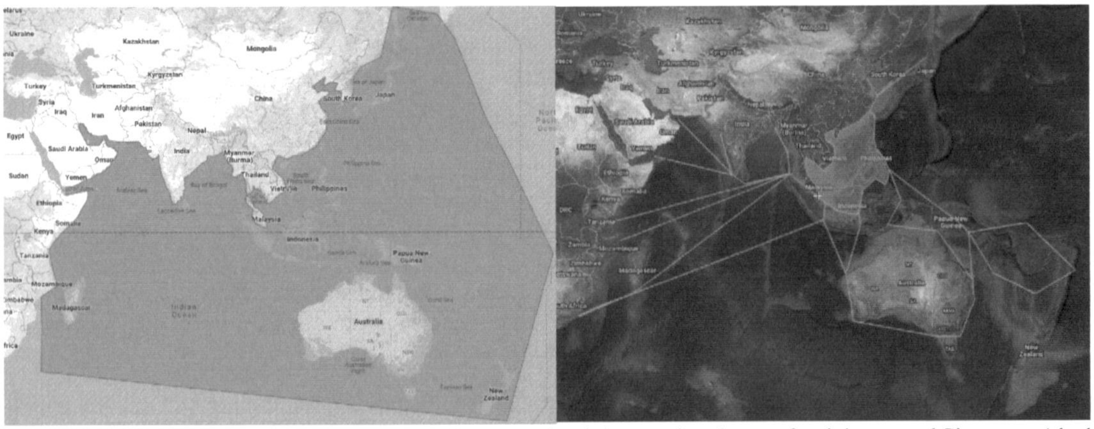

Figure 4, on the left, catchment area of the Silk Alliance; on the right, a major cluster of activity around Singapore (shaded yellow) and key routes (blue lines) of the identified baseline fleet

Figure 5 shows the heat map activity of the current baseline fleet. The baseline fleet is composed of 337 containerships across small, medium, and large size categories among the current members, with aggregated fuel demand of 3m Tonnes of HFO per year, leading to a CO2e potential reduction of 10m Tonne per year.

The Silk Alliance is now in the process of expanding its baseline to other ship types, starting with bulk carriers. The expanded baseline fleet can create a systematic change for the uptake of new fuels by concentrating effort into a shared strategy across segments.

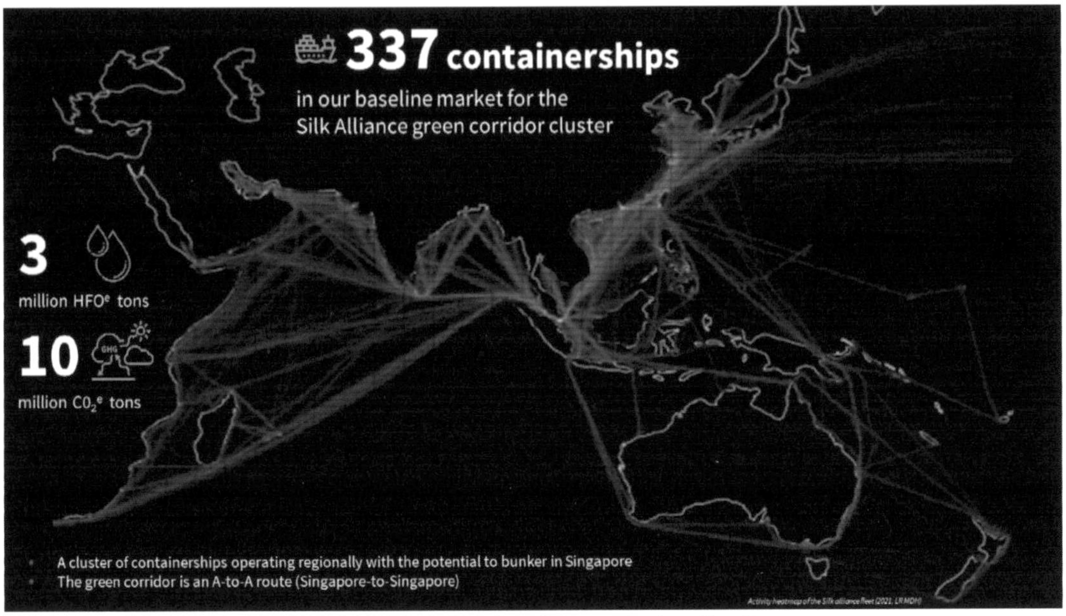

Figure 5, Heat map activity of the current baseline fleet

4.2 (c) Building a transition model

Understanding the dynamics that could influence fleet renewal is extremely important to estimate the number of newbuilds and retrofits needed in line with emission reduction goals. In doing so, estimations of costs, benefits, and investments under different scenarios can be made. Due to the large number of variables involved when assessing potential fleet renewal, a computational model that simulates the fleet transition given certain input is one of the best ways to fully understand the dynamics and risks involved in each option.

Drawing on the fleet model developed in the First Mover Framework, we have further refined the model to address the specific areas of interest of the members. In short, the model simulates the evolution of the fleet and for each year assigns fuels to each newbuild based on a pre-determined fuel strategy, while for the existing fleet of that given year, it proceeds with early scrap and newbuild replacement till a defined age limit and then it proceeds with retrofitting with certain constraints on age limit. Early scrap first and retrofitting second, are only applied if emissions goals are not met.

This model has allowed us to explore a wide range of scenarios. By settings different input assumptions, we were able to provide key insights into the questions posed.

4.2 (d) Key preliminary results

This section briefly describes the type of preliminary results of the scenario-based analysis. This is done only for a few scenarios. The interpretation of the results is not discussed here as this is not the scope of this paper.

Figure 6 shows the estimated fuel mix and fuel demand for the baseline fleet according to three different fuel transition strategies based respectively on ammonia, methanol, and methane. A distinction was made between e-fuels and re-fuels. The former indicates fuels with lower emissions factors whereas the latter, re-fuels (renewable energy-based fuels) indicates fuels with nearly zero emissions.

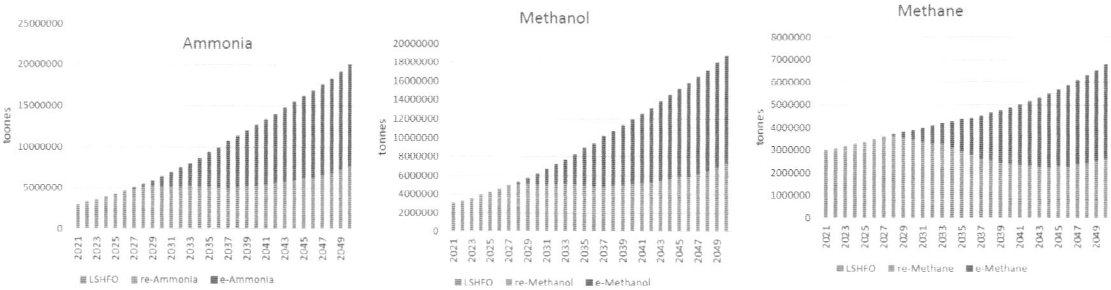

Figure 6, Estimated fuel mix and fuel demand for 3 different scenarios

Figure 7 includes two opposed scenarios that yield very similar results when normalized. In all scenarios, the carbon cost exposure in the early period of the energy transition increases rapidly, so managing and flattening the peak of the carbon costs (e.g. by increasing the adoption of renewable fuels) is one way of gaining a competitive advantage. "Pay-to-pollute" is shown to be a risky strategy – and prices will only be brought under control by switching fuel. The figures also demonstrate how the carbon cost exposure from the 2030s onward is coupled with the pace of renewable fuel uptake.

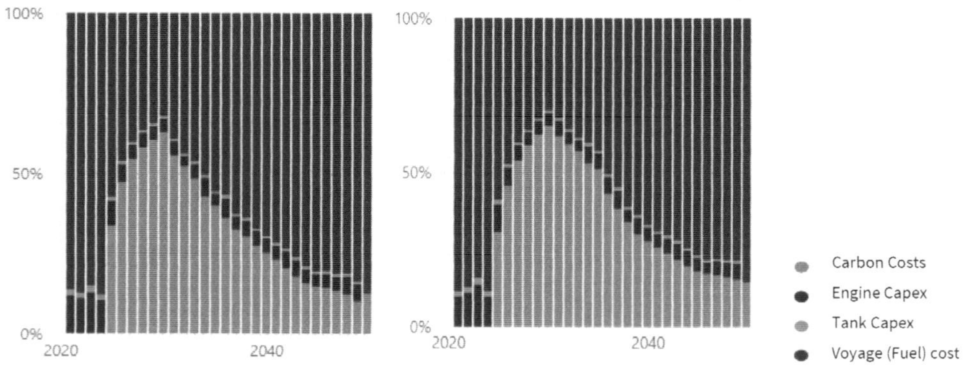

Figure 7, Breakdown of the fleet costs for the Silk Alliance's baseline fleet over time for two different scenarios. On the left, a representation of a scenario with a high carbon price and modelling a rapid uptake of renewable fuels. On the right, a representation of a scenario with a low carbon price and slow uptake of renewable fuels.

Figure 8 shows, the voyage and carbon costs for the baseline fleet calculated for each of the scenarios modelled. One can observe that there is a premium on methane fuel strategies relative to the Ammonia fuel strategies widening in the latter decades of the energy transition.

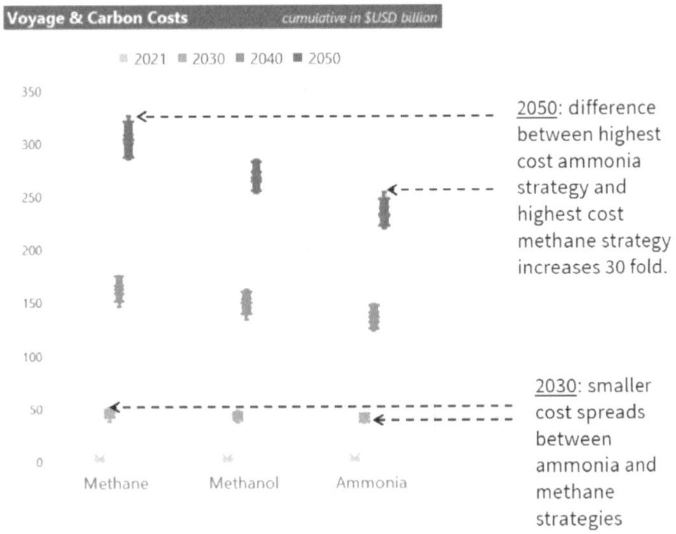

Figure 8, each candlestick represents a range of cumulative OPEX costs derived under different scenarios split by fuel transition strategy: Methane, Ammonia, and Methanol, and at different time intervals.

4.3 BUILDING AN IMPLEMENTATION PLAN

The results of the scenario analysis were used as a basis for a discussion during the third workshop. Although convergence to a single fuel strategy has not been reached, members continued to carry out this process as part of the implementation plan. We envisage that this stage may involve other analytical assessments and iterations, especially to account for the perspectives of fuel producers and governments. Figure 9 shows the initial draft used for discussion which provides the breakdown of the implementation plan highlighting the need to synchronise activities across the fleet, the fuel supply production, and the port as well as the key enablers.

Main milestones	
FLEET decarbonisation strategy	Define/agree on the strategy
	Consideration on timelines for fleet renewal, retrofit/conversion
	Forecast investment required
	Safety onboard & Forecast of the training requirements
FUEL/s production strategy	Engagement and selection
	Define/agree on the strategy
	Forecast investment required
	Consideration on timelines for fleet renewal, retrofit/conversation
PORT and bunkering infrastructure	Feasibility study (technical design)
	Consideration on timelines
	Safety at port & Forecast of the training requirements
Key enablers	Offtake agreement
	Policy incentives
	Finance mechanism
	Partnerships, bringing in new members

Figure 9, the initial template of the implementation plan used for discussion

The next steps will focus on expanding the alliance to ensure other key stakeholders are involved and designing the implementation plan.

5. INITIAL LESSONS LEARNED

This section includes several lessons learned from the work done so far by the LR Maritime Decarbonisation Hub and our engagement with the members and external actors. This initial list is mostly derived from our experience as a facilitator and from the role we have as a decarbonation hub. The section is composed of two parts. The first part includes a list of lessons and reflections on our experience in initiating and developing a green corridor project within the current landscape. The second part includes a list of lessons from the work itself undertaken so far.

5.1 THE SILK ALLIANCE AND THE CURRENT LANDSCAPE

5.1 (a) Actions and flexibility

The focus should be on actions rather than definitions – right from the start. The ultimate definition and scope will be set based on what has been achieved – so clarity and specificity on what actions are required to get the vessels on the water are needed. This may mean looking at the issues from completely different angles that are not constrained by the definitions that the process started with. For example, we may need to extend beyond the currently defined scope of shipping's aggregate demand, and instead widen the scope to capture regional demands, including energy and fuel demand from other sectors.

5.1 (b) Plan and implementation

The implementation plan until 2030 is a starting point. Concerns are raised that green corridor initiatives are only there for the mere purpose of generating an implementation plan or roadmap. This is an essential part of the process, but the effort should go also into the actual implementation and putting in place an iterative process to update the plan while the landscape changes over time. Therefore, the need for a programmatic approach with an agile implementation.

5.1 (c) The network of green corridors

While green corridor initiatives are still in their infancy, efforts have been made to create a network of green corridors to incentivise sharing and collaboration. If we accept that the focus of green corridors shouldn't be limited to the simple substitution in specific routes, and instead work more on how they may branch, pile up, merge, and contribute to systematic changes, then collaboration among green corridors should be the priority. Although each initiative has its peculiarity, and scalability potential and can be deployed in different configurations to maximise impact and accelerate the transition, we suggest that joined-up projects with other green corridors in parallel may positively determine or change the direction of green corridor initiatives. The aim of publishing details in this paper was to deliver greater transparency around the Silk Alliance green corridor cluster, and we call for more open discussion and sharing of information among other green corridors.

5.1 (d) Linkages with policy and technology developments

Green corridors are to be perceived as crucial to kick-start the transition, but they cannot do this on their own. A successful network of green corridors could create these linkages with ongoing external processes such as policy and technology developments. This remains a challenge because the effort in creating linkages is still very fragmented even among the members of a single initiative.

5.2 THE SILK ALLIANCE AND THE PROGRESS MADE

5.2 (a) Collective first movers

Green corridors are not about bringing first movers together but more about bringing key players together and making them act collectively as first movers; this is itself a challenge. Making organisations engage with competitors to become "collective first movers" remains a gradual process.

5.2 (b) An iterative process

We benefitted from bringing actors together that were a subset of the entire supply chain in the first stage of industry engagement (e.g. the Silk Alliance has not yet formally onboarded fuel producers or governments). While the theory indicates the need for system-wide collaboration, this intentional decision made the initial kick-off a manageable process. As we expand the membership, the process will become iterative because we will need to work with new and current members to revisit the first stages of the process. On one hand, the iterative process breaks down the process into manageable stages, particularly when dealing with a vast number of stakeholders across the board, and helps to refine the process with each iteration, while on the other hand, the iterative process requires a longer time to execute.

5.3 (c) Creating consensus

Moving from generalisations into "specifics" has proven to work as we have garnered interest and engagement from the members when looking at the results of an energy transition tailored to their specific fleet. However, the creation of consensus is challenging itself and consensus cannot happen particularly at this stage when the onboarding for all stakeholders has not yet happened. Workshops or roundtables are ways to drive consensus and decision-making, but due to the complexity of the problems and delivery of analysis and results in these forums, alternative means of knowledge sharing among members should be explored to supplement these efforts. In cross-industry partnerships, engagement in open floor discussion is bettered by 1-1 and smaller engagements and working groups so partners grow comfortable operating in these novel and innovative spaces.

5.4 (d) The baseline fleet and the scope
Identifying the 'baseline fleet' may change the scope of the green corridor. We benefitted from the AIS data analysis to identify the baseline fleet because it helped us to assess the activity, the wider scope, and the regional nature of the Silk alliance green corridor cluster programme. The point-to-point concept was not suitable for the identified fleet which instead acts in a relatively large cluster.

5.5 (e) Regional scope and alternative finance
The regional focus of the Silk Alliance opens the opportunity to incentivise the development of fuel supply infrastructure in strategic positions across developing countries. Governments and financial institutions should nurture not-yet profitable investments in this area as there is the expectation that they will become important for realising a just transition with societal and collective goals in the future. The investment barrier remains the largest challenge so discussions and workstreams need to be created to address this high priority area.

5. CONCLUSIONS

The topic of green shipping corridors is becoming increasingly important as many recognise the potential benefits involved in such type of initiative. There have been several announcements of green shipping corridors whose scope can vary significantly depending on the interpretation. However, we argue that the focus of these initiatives is to kickstart the uptake of alternative fuels that can generate spill-over effects for a wider sector transition whatever definition is taken into account.

The debate still focuses on defining and identifying the important elements of a green corridor and how to approach it, suggesting that green corridors remain in an infancy stage with various methodologies and approaches so far taken. However, we think that scenario analysis and feasibility studies are only the starting point to advance and establish a roadmap for green corridors The real challenge would be the actual implementation of a green corridor. It is one thing to flesh out a vision of the transition, which is technically feasible, but it is another thing entirely to put an implementation plan into practice at scale.

The Silk Alliance, a green corridors cluster, is evolving as we progress with the work. We aim on delivering and implementing a roadmap. This will require the involvement of other stakeholders that are looking at the same catchment area.

One example is energy suppliers and governments with an interest in inland fuel infrastructure with the potential to competitively produce these fuels and are within this region. We know there are different fuel candidates, however, the fuel supply infrastructure at least at the beginning could focus on the common building blocks such as hydrogen. We would welcome collaborations across different actors to create such building blocks.

Another example is financial institutions with access to the climate fund. The creation of a robust consensus across the value chain remains a challenge. So far, we noted the intention from the members of piloting the main fuel candidates and working further on potential convergence to one or more fuels, but nothing is possible without securing the financing. More effort needs to go into identifying a funding and risk-sharing mechanism to turn roadmaps and plans into actions

We adopted a system approach for this green corridor cluster. To help current and future members, we are using a framework designed with the view that both sides, the fuel supply, and the fleet should work in tandem to support the energy transition along a shipping 'corridor' to ensure its success. The framework is intended to guide but further extensive analytical assessments are essential to unlock investments. We detailed the initial lessons learned during the first cycle of the process; however, work will continue and will be completed with further iterations with new members. Simultaneously, we will look at opportunities to start parallel activities with other green corridor initiatives that may lead to synergies and further acceleration of the shipping energy transition.

6. ACKNOWLEDGEMENTS

We would like to acknowledge the support of the Lloyd's Register Foundation and Lloyd's Register for funding this work. We thank the members of the Silk Alliance for their participation and feedback in the collaborative workshops and working groups as part of the process. We also thank external organisations and the other green corridor initiatives for providing feedback to help with the refinement of the processes. Finally, we would like to thank RINA for the opportunity to present this work at the conference Scaling Decarbonisation Solutions – Reducing Emissions by 2030 Conference – Royal Institution of Naval Architects.

7. REFERENCES

1. DEPARTMENT OF TRANSPORT, U.K. 'COP26: Clydebank Declaration for green shipping corridors', November 2021, https://www.gov.uk/government/publications/cop-26-clydebank-declaration-for-green-shipping-corridors

2. GLOBAL MARITIME FORUM, 'The next wave: green corridors', November 2021, https://www.globalmaritimeforum.org/content/2021/11/The-Next-Wave-Green-Corridors.pdf .

3. U.S. DEPARTMENT OF STATE, 'Launch of the Green Shipping Challenge at COP27', November 2022, link: Launch of the Green Shipping Challenge at COP27 - United States Department of State

4. LLOYD'S REGISTER MARITIME DECARBONISATION HUB, 'First movers in shipping's decarbonisation. A framework for getting started. A case study for containership feeders in Asia, December 2021, http://mission-innovation.net/wp-content/uploads/2022/09/LR_First_movers_in_shipping_s_decarbonisation_A_framework_for_getting_.pdf

5. GLOBAL MARITIME FORUM, 'Green Corridors: Definitions and Approaches. A Discussion paper from the Global Maritime Forum', August 2022, https://www.globalmaritimeforum.org/content/2022/08/Discussion-paper_Green-Corridors-Definitions-and-Approaches.pdf

6. LLOYD'S REGISTER MARITIME DECARBONISATION HUB, THE RESILIENCE SHIFT & ARUP, 'Port energy supply for Green Shipping Corridors', January 2022, https://www.arup.com/-/media/arup/files/publications/p/port-energy-supply-for-green-shipping-corridors.pdf

7. D ENGLERT, A LOSOS, C RAUCCI, T SMITH, 'The Potential of Zero-Carbon Bunker fuels in Developing Countries. Volume 1, World Bank, Washington, DC. © World Bank. 2021, https://openknowledge.worldbank.org/handle/10986/35435 License: CC BY 3.0 IGO.

8. DECISION INTEGRITY LIMITED, 'Method for Strategic Thinking with Scenarios', online, https://www.decisionintegrity.co.uk/page41.html

8. AUTHORS BIOGRAPHY

Carlo Raucci holds the current position of Consultant at Lloyd's Register Maritime Decarbonisation Hub. He is responsible for leading the development and implementation of the hub activities for accelerating the energy transition in shipping. His previous experience includes more than 15 years of experience on projects for alternative fuels in shipping. He authored several highly influential publications on the topic of transition to zero-emissions vessels including reports on low carbon pathways which were one of the first to suggest the goal of a 50% reduction in greenhouse gas emissions from shipping by 2050, and to achieve this, zero-emission vessels would need to be in use by 2030. He holds a PhD from the University College London on the potential of hydrogen as a fuel for decarbonising shipping.

Ahila Karan holds the current position of Senior Decarbonisation Analyst at Lloyd's Register Maritime Decarbonisation Hub. She is responsible for data analysis and supporting initiatives looking to accelerate decarbonisation and uptake of alternative fuels in shipping, also co-authoring papers in this field, most recently a joint study entitled "port energy supply infrastructure for green shipping corridors". Her previous experience includes working as a research analyst for an energy trading company and has 7 years of experience covering transportation fuels.

Shane Balani holds the current position of Decarbonisation Consultant at Lloyd's Register. He is responsible for various decarbonisation projects across the value chain of shipping. His previous experience includes international exposure as a classification surveyor of new construction and existing maritime assets.

WIND PROPULSION SCALABILITY: OPPORTUNITIES & CHALLENGES

G Allwright, International Windship Association, UK

SUMMARY

Wind propulsion technologies are entering the market now as wind-assist installations and large primary wind vessels will soon be following. There are quite a number of drivers and key benefits of harnessing this free, zero-emissions energy source and these are starting to accelerate the pace of this uptake after some years of testing, validating and demonstrating the potential of this technology segment. The development of a wind-powered fleet combines three threads, tweaking the existing fleet through retrofits, transitioning to optimised newbuilds and primary wind vessels and the potential to establish a new paradigm. By collating feedback from key stakeholders in the wind propulsion field, the structure of this fleet starts to take shape and the progress in tackling the continued barriers to the diffusion and scaling of these technologies are highlighted.

1. INTRODUCTION

Human interaction with the vast oceans was made possible by the harnessing of wind power, early forays on rivers and along the coast had been powered by humans but it wasn't until wind was harnessed that vast distances could be covered. This harnessing of the wind enabled the water borne trade we know today, and oceans stopped being walls separating us and started becoming highways that brought us together - globalisation. The second energy transition was from wind power to coal and steam and then the move to oil, a transition that took over 150 years to complete, however in a commercial sense wind never went away. This free, abundant and globally available energy source has continued to be harnessed in small niche markets commercially, with dhow's still used in the Indian Ocean and down the coast of Africa and in Indonesia with the Makasa schooner's still plying their trade and small areas of the South Pacific and elsewhere. Wind propulsion technologies also continued to be harnessed extensively in the yacht, leisure sailing and racing world's leading to continued innovations and new materials.

In the late 1970s and early 1980s we saw an upsurge in the interest in bringing wind propulsion back into commercial shipping with one notable example being the 10 wind-assist vessels that were built in Japan during this period, including the 1,600dwt coastal tanker, the Shin Aitoku Maru and the 26,000-dwt bulker, the Usuki Pioneer. These efforts were driven by the energy crisis and the high cost of fuel, but due to few other drivers, these pioneering projects couldn't survive the re-emergence of cheap oil in the mid-80s.

Today however, the drivers of the re-emergence of wind propulsion as a commercial energy source for shipping are more multi-faceted and some would argue more robust. These drivers naturally include the higher price of fuel and that price volatility, but increasingly these include the need to have reliable and risk-free energy sources available to all industry segments in the economy including shipping. The other drivers are fairly obvious, including regulatory pressure applied by the need to decarbonize the sector and customer pressure coming from ESG commitments driven by social pressure and the growing concerns over climate change and the recognition of the impacts of shipping emissions on ocean biodiversity, ocean health, air pollution, human health impacts and so on. [1]

As highlighted in the recent IMO submission MEPC79 INF. 21 submitted by Finland, France, Saudi Arabia, Solomon Islands, Spain, Union of Comoros, RINA and the International Windship Association, there are a significant number of unique sales points when it comes to the deployment of direct wind propulsion technologies (WPT), these include but are not restricted to the fact that they are harnessing an inexhaustible, globally available zero-emissions energy source that requires no mining, refining, transport, bunkering or storage onboard. This energy is delivered to the point of use at no cost for the lifetime of the vessel. Wind propulsion technologies are available today, backed by a robust pipeline of systems coming through. As we can see from the table below there are significant co-benefits of wind as an energy source and those propulsion technologies are a robust, reliable and increasingly commercially attractive means of harvesting that energy source.

Table 1 – Key considerations for direct wind energy, WPT and ships [2]

Direct Wind Energy Use

Propulsion vs Fuel Efficiency	A direct propulsive energy source that can be used as a partial or main propulsion system.
Emissions	An efficient zero-emissions energy source requiring no conversion thus avoiding power loss of up to 90% for alternative fuels based on a Well-to-Wake (WtW) calculation.
Infrastructure	Requires no ongoing mining, refining, bunkering or transport – delivered directly to the point of use.
Storage	Requires no storage tanks onboard the ship, though requires installation space on deck. Wind solutions also reduce propulsion energy requirements, therefore can also reduce other energy storage systems requirements, such as batteries.
Energy Cost	Cost of direct wind energy is zero and tax free for the lifetime of the ship without competition for supply.
Energy Availability	Abundant and freely available worldwide today. Wind is however intermittent and irregular, and this variability is route specific with some routes having more wind resources than others.
Predictability	Increasingly predictable with modern forecasting technologies, a century of wind data and onboard weather stations/sensors, LiDAR etc.
Facilitation	With this free energy source integrated into an energy efficiency strategy, the use of WPT reduces the amount needed of new alternative fuels/technologies, thus helping to facilitate their deployment in a cost-efficient way.
Wind Propulsion Technologies & Ships	
Wind-assist	5-20% of the propulsive energy requirement delivered as a retrofit based on motor ship profiles with the potential to optimise these operationally up to 30%.
Primary Wind	50%+ of propulsive energy requirement (potential for 100%) especially on newbuild ships & using wind optimisation techniques.
Viability	WPT are already demonstrating their viability as a technology on 21 vessels (25 vessels by end of Q4 2022) across six fleet segments.
Safety	All systems and ship designs are certified by class and comply with SOLAS/COLREGS. WPT have the potential to also add to safety as a secondary propulsion system in case of engine failure in certain conditions. Certain wind systems can also improve the stability and comfort onboard, as they can dampen rolling, etc.
Operations	Air draft and port operations are key issues solved with folding/retracting/movable systems + modular possible.
Availability	Available now with 10+ companies supplying varied size/power WPT already into the market. 10+ more are in pre-market, 20+ in R&D pipeline.
Compatibility	WPT are compatible with all other fuel and technology choices and various types can be deployed in all fleet segments.
Integration	Integrating WPT into Energy Management Systems (EMS) will further optimise systems and fuel savings.
Speed	While speed is primarily determined by market factors, if ships do reduce speed, then the ratio of the power delivered by WPT rises. However, WPT can also be used to maintain higher speeds with lower fuel consumption.
Routing for Wind	Increasingly sophisticated wind-routing software is available, enabling operators to maximise wind/minimise fuel and/or maintain schedules.
Range	Using this free energy source, WPTs can increase ship range for the same fuel usage.
Onboard Energy Production	Surplus wind power can be harvested and used to produce electricity/alternative fuels through onboard wind turbines or by other means.
Autonomous Shipping	Wind gives a large degree of energy autonomy to ships aligning with development of autonomous ships.
Lifecycle Analysis	The energy used in the production and operation of the WPT is minimal and recouped quickly. Materials will increase in recyclability as production processes develop.

Training & Crew	Highly automated + Remote sensors help with predictive maintenance. No extra crew required but some training is necessary.
Underwater Sound	Additional benefit of WPT is the reduction of underwater noise with the option to be fully silent when in sensitive marine areas.

For the purposes of this paper, there is a need to differentiate between 'wind-assist' systems, where the installation of the WPT is deployed in support of the primary propulsion system and more substantial primary wind vessels, where a new build vessel is designed to have WPT as the primary propulsion system backed up by engines that are used for manoeuvring, motoring in very heavy weather or during periods of very low or no wind.

2. WHAT WILL A WIND PROPULSION ORIENTEDFLEET LOOK LIKE?

It is extremely challenging to forecast with any accuracy what the fleet will look like in 2030, let alone 2050 or beyond, however there are some factors and drivers that are likely to remain constant or that will continue to be key going forward. With the use of wind propulsion there are a number of constants, the fact that the energy will remain free and not subject to the vagaries of volatile energy markets and supply constraints and that there will never be a need for land-based infrastructure to produce and deliver that energy directly to the vessel. These two factors alone are very significant in any assessment of the broader economic framework but also for the future fleet and how it operates. We also have the strong likelihood that wind propulsion systems and their support technologies will continue to be optimised, be increasingly well-integrated into vessel operating and energy management systems and the baseline costs of the systems will continue to fall in line with a learning curve estimated to be around 10%, similar to that of offshore wind, thus every doubling of installations will deliver a 10% reduction in cost. [3], [4]

The scale and power of systems is also likely to increase and their deployment as primary power sources on larger vessels will increasingly enter the market from 2024/25 onwards. This trend is already apparent when we observe the rotor sail market, where initial installations were limited to options of fixed 18-24m rotor sails, while that offering has now increased to 30-35m options with rail systems and hinged, retractable systems now deployed in the market.

On the fuel, pricing and regulatory side of the equation, the momentum is clearly towards higher prices for fuel, especially when carbon pricing is included for fossil fuels from the EU ETS and subsequent carbon levies imposed regionally and at an IMO level [5]. Alternative fuels such as ammonia, methanol and hydrogen will likely enter the market gradually, and initially be subject to carbon taxes if those are produced from fossil fuels. The move towards the 'green' version of these fuels will increase the cost for these fuels anywhere between three- and five-times existing fuel prices, for example the projection of $2,000/ton for green methanol to fuel Maersk's order for methanol powered container vessels [6] against current VLSFO prices in Rotterdam of $635/ton (12 Nov 2022)

Regulatory drivers are also likely to strengthen, with IMO regulations around the next phases of EEDI, the introduction of EEXI in 2023 along with a progressively strengthening CII. These will be supplemented and pressured further by regional legislation such as the FuelEU Maritime and other elements of the EU 'Fitfor55' program. Therefore, the momentum is building for higher energy prices and increased climate focused restrictions on emissions from vessels.

One other variable is the shifting of trading patterns in the future; which commodities will be transported? Will energy still be tankered at scale? From where to where? This uncertainty should push the industry towards a position of building a future fleet that is more resilient to change, with more flexibility of size/type/energy for vessels to deal with these shifts in global trading activity etc., however this has not necessarily been the case in the past. [7]

So, what does this mean for wind propulsion? Well, as these trends strengthen and wind gains credibility and validation from the demonstrator vessels in operation, we could see the growth of wind-powered vessels follow three general pathways:

(i)Tweaking - the tweaking of the existing system, where retrofitting the fleet and optimising vessel operations delivers significant savings of 5-20% and potentially up to 30% from ships operating within the existing motor vessel paradigm. This thread is already well underway as is evidenced by over twenty retrofit and newbuild wind-assist vessels in operation. New modular technologies could further boost this uptake by encouraging pay-as-you-sail business and finance models, where the costs are moved from CAPEX to OPEX. [8].

(ii)Transition – In parallel, there is a transitional thread developing, where newbuild optimised wind-assist vessels are entering the fleet en-masse and a sizeable number of primary wind designs being built utilising extensive routing and speed adjustments to deliver 50-90% reductions in fuel use on selected routes. This thread is only in its foundation period, however the period 2023-25 will see up to five large primary wind newbuild and optimised wind-assist vessels entering operation. This thread could be further accelerated by the adoption of new approaches such as mandating all new build vessels to be made wind-ready and through the cargo owner specification for primary wind vessel adoption and long-term contracts to assist with this, such as the recent tender awarded by Shippers' Coalition for Low Carbon Maritime Transport, under the aegis of France Supply Chain and the Association of Freight Transport Users (AUTF). for a twice weekly transatlantic route between Europe and North America, operated from 2024 onwards by container ships powered mainly by wind which was awarded to Zéphyr et Borée and their proposed fleet of 10-12 wind-powered 600TEU container vessels. [9]

(iii) Transformation - Finally, a more disruptive thread could develop, with a transformative, total overhaul or replacement of the way shipping operates and the ships themselves. The primary wind vessel becomes the standard vessel type across industry segments, operational changes are standardised to take into account wind propulsion incorporating 'trade wind' corridors and weather routing etc. Energy harvesting systems would be deployed on these ships that enable the capture of excess wind energy to generate onboard fuels for use when there is no wind or for manoeuvring and even energy ships that produce a tradable cargo of green alternative fuel. [10]

3. WHAT IS HAPPENING TODAY?

Wind propulsion, and currently that primarily means the installation of wind-assist solutions, continues to gain traction with installations on 22 large vessels (over 400 GRT) and these will likely top 25 large vessels by the end of 2022, with a combined dead weight tonnage of 1.7 million dwt. This number of vessels is likely to double by the end of 2023, with an additional 24 vessels scheduled for delivery next year, exceeding 3.3 million dwt (taking into account potential logistics and yard delays).

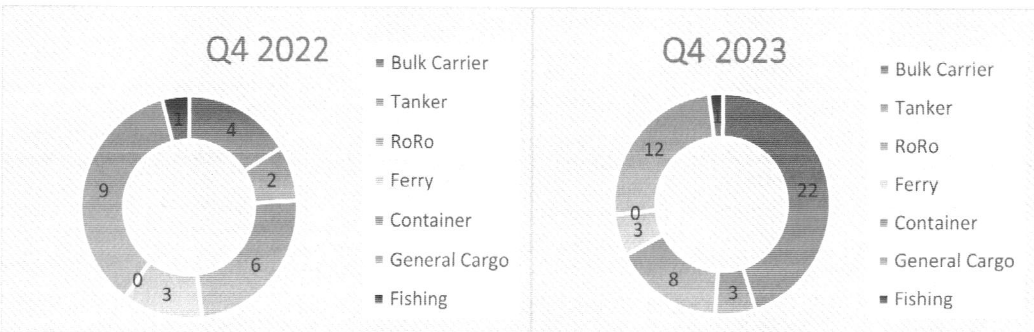

Fig 1. WPT installations by fleet category 2022-23

This total does not include the number of small cruise vessels in operation using traditional sail, which currently number around ten ships with an overall tonnage of approximately 50,000GRT. There are also a growing number of smaller vessels, with traditional sail cargo ships (under 400GRT) and retrofits on small fishing ships.[11]

These numbers however don't tell the whole picture, as the growth pattern of installations has been patchy to date, but the upward trend has strengthened substantially in 2021-22. The recent pattern indicates a substantial growth in installations going forwards, whereas the first 25 vessel installations have been spread over a period of 12 years from 2010 with the launch of the first large commercial wind-assist ship for decades, the four rotor sail equipped E-ship 1. The following 24 vessels will all be installed in the space of 12 months. This growth in installations has been gradually picking up pace, though distorted somewhat by Covid restrictions and logistics challenges. If account is taken for the Covid pandemic and its direct effect on delaying installations, 2020 should have had five vessels installed, 2021 then would have had seven installations and 2022 will see at least nine as opposed to three, nine and nine respectively.

The breakdown of retrofit and newbuild activity is also of interest, with the period from 2010-2020 having a 4/5 split between newbuilds and retrofits, then in 2021 shifting to 1/7 split, this mostly retrofit program in part reflecting a cautious approach by multiple shipowners (and the effect of the EU WASP project) in validating and trialling the technologies. That trend has again swung back to a 4/5 split of new build vs retrofits in 2022 indicating that both markets are gaining traction and possibly indicating an increased interest going forward for primary wind projects, with TOWT having started the build of two 88m, 1,100dwt general cargo ships for delivery by the end of 2023.

4. WHAT WILL THE MARKET LOOK LIKE IN THE FUTURE?

These figures and short-term projections feed into and align quite well with the large ship deployment forecasts that have been made through EU and UK government research. The EU commissioned report 'Study on the analysis of market potentials and market barriers for wind propulsion technologies for ships' [12] headlined that the maximum market potential primarily for bulk carriers and tankers would range between 3,700-10,700 installed systems by 2030 depending on bunker fuel price, speed of the ships, and discount rate applied. To reach these numbers in those two sectors would require roughly a doubling of installations year on year, and as noted in the previous section, that trend is starting to develop. The EU report however covered only three fleet segments and assessed only a small selection of WPT (four rig systems) with limited validated data, thus the growing segments of RoRo, general cargo and RoPax were excluded as were the many variant rig designs and application. This forecast also was made prior to the IMO Initial Strategy of Decarbonisation (IMO 2018) and didn't take into consideration any additional regulatory or carbon pricing effects.

In the report, the diffusion modelling results (fuel price of US$450/tn) shows the uptake of these four WPT systems in the two most applicable fleet segments (tankers and bulkers) identified in the report. This naturally does not take into account recent Covid pandemic and conflict challenges and disruption. The model does however give a clear outline for the diffusion pathway for the technology (likely with a 12-18-month lag to the numbers outlined in the table below)

4.1 HYDROGEN FUEL CELL SIMULATION

The results for the simulation based on a setup of 40 MW hydrogen-fed PEM fuel cells and 10 MWh of Lithium-ion batteries are shown in Figure 4, with the latter including the state of charge of the batteries. The energy system was set up to run the fuel cells at a constant output for as long as possible, but they have also been ramped up or down accordingly to avoid overcharging or undercharging the batteries.

Ship type	Build Type	2020	2021	2022	2023	2024	2025	2026	2027	2028	2029	2030
Tanker (5,000-120,000 dwt)	Fleet	2,892	2,915	2,938	2,961	2,984	3,008	3,022	3,036	3,050	3,064	3,078
	New build with sail	0	3	14	64	70	107	120	154	160	169	175
	Retrofit with sail	0	0	23	122	199	201	201	202	203	204	205
Bulker (0-100,000 dwt)	Fleet	10,718	11,231	11,743	12,256	12,768	13,281	13,914	14,547	15,180	15,813	16,446
	New build with sail	0	8	31	166	339	528	620	632	631	641	575
	Retrofit with sail	0	6	23	122	426	443	464	485	506	527	548

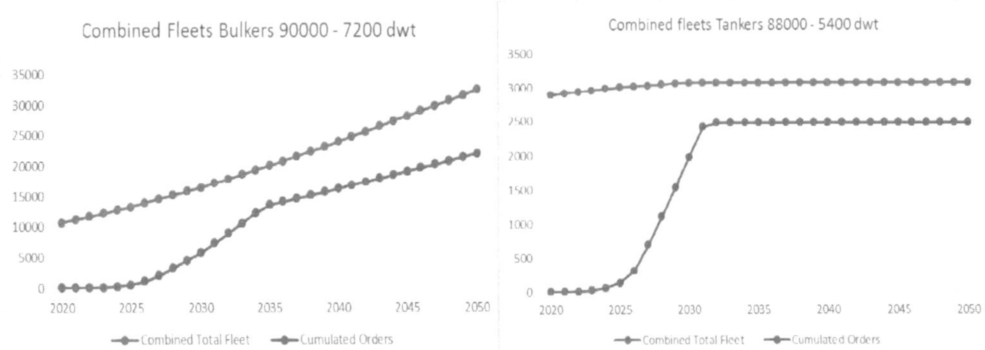

Fig 2. Sensitivity Analysis: Slow speed, oil price $450, discount rate 5% - orders of WPT ships to 2030 [12]

Whilst these projections provide an approximation of the uptake of WPT, they follow a relatively linear development curve, whereas in reality these numbers are likely to follow more erratic patterns, reacting to build and yard cycles, fuel price surges and regulatory compliance.

Projecting further out, the UK Government's Clean Maritime Plan (July 2019), sees the total market for wind propulsion growing to £2 billion per year, with the research backing the plan (p43-44) forecasting the market size by 2050 (assuming 50-100% abatement of GHG) for wind propulsion would be 37,000 - 40,000 ships or 40-45% of the global fleet. [13]

In the small ship segment, the largest group of small ships in operation that can benefit from WPT are the 2.86 million motorised fishing ships (63% of the total), however currently there are only a small number of projects in this segment in part due to financial constraints. There are a number of projects in the pipeline, including a German government financed sail-cargo demonstrator in the Marshall Islands that should be deployed by the end of 2023 or early 2024 and which may act as a template for a scaled fleet of these vessels to service SIDS in Pacific and beyond. [14] There are also 64,000 larger fishing ships over 400GRT. These fishing vessels have special considerations, usually with limited working area on deck, challenging operational profiles and stability limitations, however ships with long voyage times could make substantial fuel savings from the use of WPT. Additional research and demonstrator ships are required to assess and validate these installations. [15]

These figures estimate both retrofit and newbuild penetration into the existing fleet structure, however they adopt a rather static or narrow view of the wind propulsion technology rollout, with little regard shown for the scaling of the power capacity of the systems and they don't assess the dissemination of these technologies in scenarios where there are additional stringent regulatory drivers or substantial fuel supply or price volatility. The quickening rise in installations this year and the capacity for deployment across the fleet and growing mainstreaming of WPT has taken many of the decarbonisation pathway authors by surprise, including the scenarios outlined in the IMO 4th GHG study [16], which failed to assess all WPT in the market, their capacity to be scaled and their applicability across the world fleet, nor that even the container sector is now looking at wind options for newbuild vessels, albeit at the small feeder vessel level to date.

5. WHAT HAS BEEN HOLDING BACK THE WPT UPTAKE?

The re-emergence of WPT and their uptake in the market has naturally brought with them their own unique set of barriers and factors that create diffusion drag, however overall, the scaling of WPT will likely follow the relatively standard diffusion pathway for heavy marine equipment. [17] The key barriers identified in the EU-commissioned report were:

1. (Trusted) information on the performance, operability, safety, durability, and economic implications.
2. Access capital for WPT development, especially building/testing of full-scale demonstrators.
3. Incentives to improve energy efficiency/reduce CO2 emissions of ships.

The report noted that; "These key barriers are interrelated, with the most crucial interaction being a chicken-and-egg problem between 1 & 2. In order to breach this, we see the development of a standardized method to assess WPT combined with test cases to develop this assessment method as the most important starting point for overcoming the barriers."

The barriers were revisited during a workshop involving over 40 key stakeholders from the industry along with WPT experts [18] and the summary report while there had been significant since 2016 there was still continued work required on these key issues and the wider challenges as can be seen in the table below. To quickly summarise the areas of most concern, the workshop adopted a traffic light appraisal: [Red – No action taken, Red/Amber – limited action underway, Amber – action solidly underway but not completed, Green – completed/effectively solved]

Table 2. Summary findings of the Wind Propulsion Strategy Workshop, June 2021

Challenges	**Development 2017-2021**	**2022-23 Outlook**
Technical Issues		
Diffusion of Knowhow	[Red ➡ Red/Amber]	Increasing but need to move faster/wider
Class Guidelines for WPT	[Red/Amber ➡ Amber]	Need primary wind guidelines
R&D Pipeline of WPT	[Holding at Amber]	Needs additional work
Policy, Regulation & External Issues		
Existing Regulations to Reduce GHG	[Holding at Red/Amber]	Increased movement at IMO, EU and national levels

Incentives for Energy Efficiency/Reduce GHG	[Holding at Red]	Improvement with movement at IMO, EU and national levels
Inclusion of WPT in Policy Pathways & Regulations	[Red ➡ Red/Amber]	Some improvement with movement at IMO & EU levels
Business, Operations & Finance Issues		
Availability/Access to Capital for R&D, Pilots & Installations	[Red/Amber ➡ Amber]	Improvements in R&D funding & some in installation funding
Diffusion of Demonstrators & Knowledge of WPT	[Red ➡ Amber]	Increasing number of WPT & ships aligned with forecasts
Trusted 3rd Party Performance Information	[Holding at Red]	Some improvement but requires more development
Fuel Prices & ROI	[Red ➡ Red/Amber]	Higher fuel prices and carbon pricing under consideration.
Commercial Agreements/Legal	[Red ➡ Amber]	Additional work required
Scepticism/Perception Issues	[Holding at Red/Amber]	Gradual improvement

This assessment was then followed by a more detailed summary of these barriers and potential action required, incorporating many of the issues discussed in depth during the multi-stakeholder workshop by the WASP project publishing these aligned results in December 2021. [19] The IWSA membership survey held in May 2022 highlighted that the R&D pipeline has continued to strengthen however whilst R&D funding is coming into the sector and benefiting those OEM's already operating in the market, there is still a significant challenge in accessing pre-market funding and enabling pilot/demonstrator roll-out.

The regulatory framework has also continued to develop positively and IWSA members across the board have noted that the upcoming adoption of EEXI and CII in 2023 are providing positive decarbonisation signals to the industry, however the low level of ambition and the need for additional teeth for these regulations have been frequently cited. The next phase of EEDI is also seen as a driver for newbuild wind propulsion projects. While the adoption of the IMO MEPC.1/Circ.896, Guidance on Treatment of Innovative Energy Efficiency Technologies for Calculation and Verification of the Attained EEDI and EEXI in December 2021 [20] has updated and reinforced the assessment of wind-assist installations, this circular still requires modification and upgrading to enable a level playing field for all WPT variants and also to ensure that more substantial wind-assist and primary wind systems are correctly assessed. Additional regional regulation such as FuelEU Maritime are also contributors, though wind propulsion assessments are still judged as 'energy efficiency 'measures rather than 'propulsion energy' providers in their own right. This misrepresentation of direct wind propulsion leads to a fuel centric approach that focuses on regulatory and financial levers that emphasise and promote commoditised fuel structures rather than a wider total energy perspective and will, if not remedied, likely continue to slow diffusion of WPT, especially where that is a primary wind application.

Fuel prices and fuel price volatility has also seen a significant shift. The war in Ukraine and ensuing energy crisis in Europe and elsewhere has generated a lot of uncertainty in the market. Increased prices and the high level of volatility also plays significantly into the business case for WPT where the energy supply isn't connected to energy markets and the energy cost itself remains unchanged at zero throughout the life of the ship.

The combination of this high-level workshop, the detailed report and survey outlined above have helped inform much of the work underway, as listed in the table below which is a non-exhaustive list of some of those programs:

Table 3. List of some key WPT programs and projects tackling barriers currently underway

Program	Duration	Main Objectives	Partners & Website
Wind Assist Ship Propulsion (WASP)	2019-2023	Research, trial and validate the operational performance of a selection of WPT solutions	14 partners – univs, shipowners etc, in the North Sea area (EU project) https://northsearegion.eu/wasp/

WiSP JIP: Phase I & II:	2019-20 2021-23	Improve methods for transparent performance prediction, provide ship owners/operators with fast low-cost predictions for their fleet and to review the regulatory perspective including status of rules and regulations, identify gaps and make recommendations, and provide examples on establishing compliance.	20 industry partners including OEMs, shipowners, class & R&D centres https://www.marin.nl/en/jips/wisp-2
International Towing Tank Conference (ITTC)	2022-2024	Definitions of technologies, review methods of ship model hydrodynamic tests, wind tunnel tests, CFD, ship dynamics simulations and routing & wind/wave data. Derive KPI's for performance evaluation, review safety and regulatory issues & derive WPT sea trial standards.	Specialist Committee on Performance of Wind Powered and Wind Assisted Ships https://ittc.info/media/10159/tor-revised-after-conference.pdf (page 14)
The European Sustainable Shipping Forum (ESSF)	Apr-Dec 2022	Evaluate state-of-art WPT & market development. Assess EU regulatory impacts & make recommendations + WPT technical validation.	Ship Energy Efficiency Sub-Group – wind propulsion workstream
Optiwise	2022-25	Extensive simulations including aerodynamics, hydrodynamics, routing and energy management holistically brought together. Assess manoeuvring/seakeeping + Deliver open guidelines for integrated system optimisation with WPT and smart measurement/control.	11 industry partners (EU project) https://www.marin.nl/en/jips/optiwise
International Windship Association (IWSA)	Ongoing	Knowledge, policy & project/network support for all of these projects (and other more specific technical, research and training projects)	www.decadeofwindpropulsion.org (campaign site) www.wind-ship.org (main site)

6. ARE THERE SIGNS OF SCALING AND MATURITY IN THE MARKET?

As we have seen in the previous section on market developments there are significant opportunities to be had, however these are predicated on the continuation of the initial market uptake shown previously in Fig 1 and this will require a significant level of scaling within the market.

Through the IWSA membership survey in May 2022 and extensive interviews with members, including the IWSA 'Listening to the Wind' interview series and other podcasts we are getting a far clearer picture of a growing number of indicators pointing towards the foundations being laid for substantial scaling of activity and early signs of maturing:

- Availability of robust and validated technologies – there are currently more than ten OEMs with wind propulsion systems in the market either as fully validated products or in the process of commercial operation validation.

- New R&D entrants – the number of projects and technologies that are entering and accelerating through the R&D process has been growing steadily over the last two years, increasingly at roughly 10% per year and this trend is expected to continue.

- Rate of installations– as already noted, there will be a significant increase in the rate of installations in 2023, effectively doubling the number of installations made in the fleet to date (since 2010) in the space of 12 months. This will double the number of ships and more than double the number of rigs installed, bringing the dead weight tonnage under wind-assist to over three million dwt.

- Installation size & variants – in the past two years there has been a marked increase in the sizes and variants of WPT's being offered or developed. Rotor sails are now available up to 35m in height and we have seen the two new hinged rotor installations in this period. Containerised, flatrack suction wing and foil systems have been developed.

- Primary wind project pipeline – this is a newbuild segment thus has taken longer to secure funding and build contracts, however this segment is starting to gain momentum with projects in Sweden, Costa Rica, Marshall Islands and France that have secured contracts and steel already being cut on the first two of a fleet of 1100dwt primary wind general cargo vessels.

- Customer demand patterns shifting – there has been a growing engagement from cargo owners and charterers over the last two years, including Cargill, Vale, Michelin, Marubeni, Ariane Group, Enviva, Renault Group etc.

- Partnerships and collaborations – an increase in collaborations with industrial partners including companies such as Yara International, Alfa Laval, CGM-CMA, etc. alongside agreements with shipyards and aftersales service, for example with Wartsila.

- Production base investment and relocation – a number of the front running European OEM's have started production lines in China where costs are lower and are in close proximity to large shipyards that are increasingly adding WPT's to standard equipment lists. Additional investments are also being made in Europe.

- Geographic diffusion – there is a growing number of OEMs, shipowners, yards and suppliers from across Asia and the US exploring and engaging with WPT development thus broadening the geographical scope outside of the traditionally strong European region.

- Classification – most of the main Class societies now have WPT guidelines and are gaining increased experience in dealing with WPT installations and enquiries. IWSA has recently been joined by two additional class societies bringing that number to five and the aforementioned guidelines are being updated and upgraded, most recently by DNV. The Russian class society is the latest to release a set of WPT guidelines.

Taken individually no one of these categories is a firm indicator of scaling and the start of maturing in the sector, however taken collectively these create a strong argument that the foundations for that process are in place and the sector is already attracting hundreds of millions in investment.

As the effect of these grows, so they feed into the 'learning rate', which will become increasingly evident. The CE Delft report adopted a 10% learning rate for each doubling of installed capacity (usually 20% for technology efficiency), with WPT quite a diverse technology segment and the shipping sector is complex and 10% is closer to the level observed generally in marine equipment and also the offshore wind sector that is somewhat comparable.

If we follow this learning curve based on the growth projections indicated just in the bulker and tanker sectors, then we will likely see production costs halved by 2026/7, if not earlier, with prices for those systems falling in alignment with that. If this cycle continues at a similar pace, then it will approach halving again by the early 2030's after which it naturally slows. The economies of scale and increased competition as more WPT systems become available and validated will substantially increase diffusion, but this is also likely to encourage larger and more efficient system deployment.

Additional market opportunities are evident in a number of development trends. In the technology field, there is modularisation, with containerisation or other options that allow for systems to be swapped in/out, which could also lend itself to new business models of leasing or pay-as-you-save which have the potential for very rapid diffusion of technology where the majority of the heavy CAPEX is moved to OPEX, paid for by the reduced fuel required to run the vessel. There is also the potential around the use of a 'wind-ready' approach in newbuilding where vessels receive structural reinforcement and other design aspects are adjusted during the build in anticipation of later retrofitting WPTs. Primary wind ship designs that have energy harvesting technology onboard to harvest excess wind energy to produce electricity or alternative fuels on board as fuel or as cargo are under development and are already deployed on a small scale aboard the research vessel the Ocean Observer and the mega yacht Black Pearl as examples.

7. CASE STUDIES

7.1 WIND-ASSIST VESSELS

The wind propulsion sector and the technologies are in the early stages of deployment and published studies of validated performance results are still scarce and need to be assessed carefully as they are subject to a large number of variables. A simulated study of a 19,500dwt asphalt bulker on a North Sea route with three rotors installed have shown savings of 21.7% on the straight or shortest distance route and reduction on the fuel optimised route of 29.5%. [21]

Equipment costs at the time and a assuming a 5% discount rate and with a fuel cost of \$550/t the ROI was found to be 13.1 and 9.7 years respectively, however if we assume an average fuel cost for 2022 of \$750/t (6-month average at time of writing) then ROI are 8 and 6.5 years respectively. With a carbon price added at \$80/t (\$240-250t fuel)– then ROIs for the fuel on optimised routes comes down to 4 years, higher carbon pricing naturally reduces this further.
These calculations utilised routing optimisation, however, speed was kept steady to reduce significant schedule/utilisation changes on the fuel optimised route, however speed-optimisation would also likely lead to significant improvements.

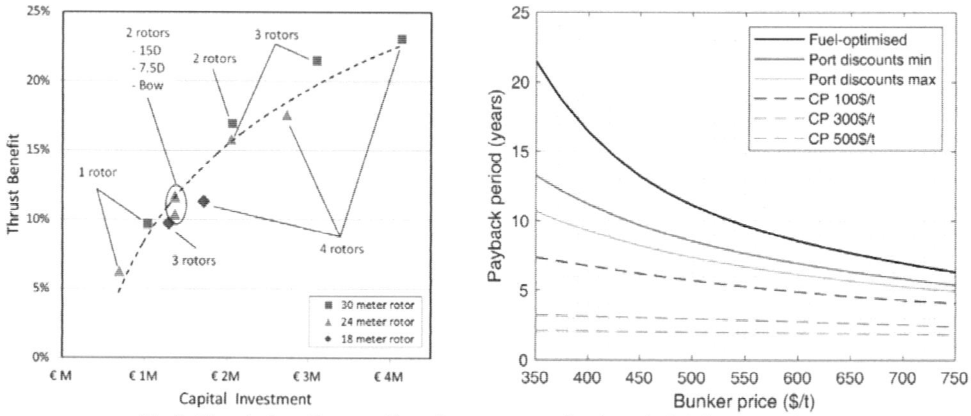

Fig 3. Simulation: Rotor sail performance, payback period & bunker price [21].

To support these calculations, additional recent simulations based on three different routes and vessels show savings per rotor of 3.4%, 5.6% and 7.2% savings, accounting for 10-28% savings in fuel without speed or routing considerations. [22]
The costs of the installations in this simulation are based on 2019 prices, and as installations increase, it has been estimated that wind propulsion systems will likely follow the standard learning curve of maritime equipment, approximately 10%, thus as installations double the technology costs reduce by 10% and this is likely to be reflected in the market price, especially as price competition increases with scaling and new entrants, thus potentially halving production costs by 2026/7.
This study's results have been further backed by a recent conference paper presented by NAPA and Norsepower at RINA's International Conference on Computer Applications in Shipbuilding 13-15 September 2022, that found similar results from 3 x 35m rotor sails installed on a VLCC operating on cross-ocean routes with voyage optimisation adding significantly to those savings, as shown in the figure below. Of course, voyage optimisation is also applicable and effective for vessels without sail installations on standard routes. [23]
In January 2021, two 35x5m Norsepower tiltable Rotor Sails were installed onboard the roro vessel, SC Connector owned by Norwegian shipowner SEA-CARGO operating in the North Sea market. According to Norsepower sources, the two Rotor Sails on the SC Connector have generated a maximum thrust force equivalent of 7 MW propeller shaft power during testing. Furthermore, the top speed peaked at 19.6 knots. The peak speed was achieved by rotor sail power (estimated at 7 MW) and approximately 3 MW from the main engine, demonstrating that the system generated more power than the installed main engine.

Picture 1. SC Connector (Courtesy of IWSA)

Additional benefits have also been noted, where seakeeping ability of the ship was considerably better by reducing the roll and by enabling higher speed in harsh weather. The crew have also learned to take the wind forecast into account when planning voyages over the North Sea. The ship has demonstrated that wind propulsion when combined with voyage optimization and route planning can provide even higher savings than with wind propulsion alone.

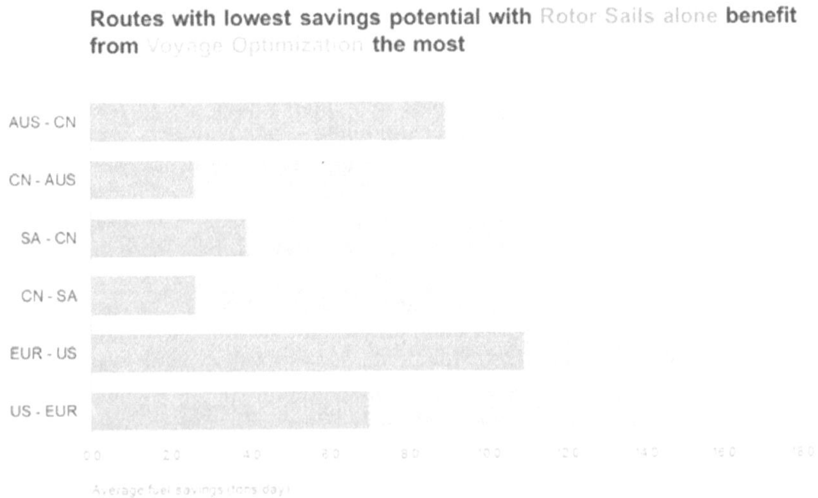

Fig 4. Routes with lowest savings potential with Rotor Sails alone benefit from voyage optimisation the most. [23]

Additional validation of retrofitted WPT has been underway using various devices on five different general cargo and RoPax vessels as part of the WASP project, an EU funded project that has installed suction wing, rotor sail and rigid wing sail systems. While the data generated from the project is valuable in assessing performance on these vessels, it is of course challenging to use this comparatively as each of the installations varies in size and type of vessel, size and power of the installed devices and the routes and timing of the voyages. However, these installations have provided a significant amount of performance data, with variations on routes providing up to 12% of fuel saving. A number of the speed trial results have been released confirming average savings for the vessels of up to 5% in operation without voyage optimisation or modifications made to the motor vessel operational profiles of the vessels. [24]

7.2 PRIMARY WIND VESSELS

Picture 2. Oceanbird design 2020 (Courtesy of IWSA)

A case study done for a primary wind vessel was carried out for the wPCC, the forerunner design of the AlfaWall Oceanbird in 2020. This 210 metre, 13,000GRT Car Carrier has been designed by Wallenius Marine and the project is now designated as the Oceanbird design, being taken forward by the AlfaWall collaboration with Alfa Laval. While this design has seen a number of iterations the projections made in 2020 are illustrative of the potential for a primary wind vessel of this size and type. The need to reduce speed naturally lengthens the duration of voyages, but also could deliver up to 90% reductions in fuel and attributed emissions for windy North Atlantic routes. To accurately compare the wPCC to a car carrier that does not sail, a preliminary power-speed curve was obtained from SSPA to determine the power needed for the hull at 8.0 and 11.4 knots, the wPCC's two average speeds. (data excluding resistance from weather). Average propulsive load @ 11.4 knots = 2,680 kW . Average propulsive load @ 8.0 knots = 992 kW

Using a Total Cost of Ownership (TOC) model, the wPCC's propulsion system ranges between €48.9m and €128.1m. The cost of a ship using BioMeOH and reaching a similar climate impact ranges between €159.7m and €1,268.6m. In the scenario using the lowest speed of 8.0 knots, when discounted cash flow has been applied and conditions are most favourable to the ships without sails, the ship using BioMeOH is 2.5 times as expensive as the wPCC which is operating under its most unfavourable conditions. In the same scenario the ship using LBG is 2.6 times more expensive but with a climate impact at least five times as high. Furthermore, this is also the scenario where the LNG ship and wPCC is closest in total ownership cost and differs by 68% in favour of the wPCC. Without applying discounted cash flow, the ship using BioMeOH is at least three times as expensive as the wPCC. [25]
On a final note, when it comes to primary wind vessels and the move into the containership market then there has been significant developments in this segment with two projects underway involving Zephyr Boree and Michelin. Michelin already has an agreement with the new build RoRo vessels to be built by Neoline to transport some of their finished products on North Atlantic routes at up to 80% reductions in fuel/emissions, however, as previously noted, the recent awarding of a tender for a twice weekly, primary wind transatlantic route won by Zéphyr et Borée will see a fleet of 10-12 small new build primary wind propulsion container vessels, each with a 600TEU capacity. The company has also received an AiP for a larger 1,800 TEU wind powered vessel too. [26]

7.3 VOYAGE OPTIMSATION

As noted above, voyage optimisation can lead to significant additional benefits. Using a wind-assist model of a Panamax bulk carrier with 4 x 35m rotors, [27] a series of route and speed optimisation simulations help to indicate the level of additional savings that can be accrued, and these correlate with other studies/simulations in the table below. When speed is reduced along with routing optimised for wind then far higher results are indicated in the 3-D optimisation curve, dependant on the voyage increase time selected, of between 20-45%.

Flettner rotors on a similar ship size

Author	Year	Ship size	Savings	Speed	Route	Flettners
Lu et al.	2020	80-120,000 DWT	6.5-8.9%	16 knots	Atlantic	One 18x3m
Paakkari et al.	2020	60-100,000 DWT	13%	11.9 knots	Global	Four 30x5m
Comer et al.	2019	110,000 DWT	1.8%-4.7%	Profile	Equatorial	Two 30x5m
Nelissen et al.	2016	90,000 DWT	17-23%	12.3-10.5 knots	Global	Three 48x6m
Howett et al.	2015	80-120,000 DWT	21-24%	12 knots	Pacific + Coastal	Four 39.4x6.4m
Traut et al.	2014	50,000 DWT	5%	11.2 knots	Atlantic	One 35x5m
Mason et al.	2021	80,000 DWT	12.2-25.2%	12-8 knots	North Atlantic	Four 35x5m

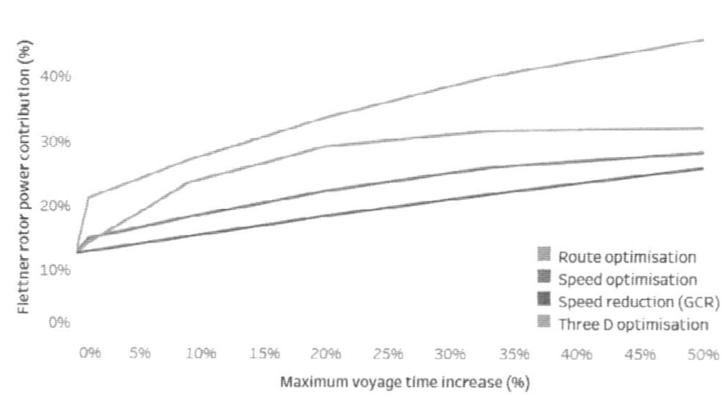

Figure 5.14: Flettner rotor power contribution for all four simulation types: route optimisation, speed optimisation, three-dimensional optimisation and speed reduction.

Fig 5. Break down of benefits from voyage optimisation [27]

These results in Fig 5. are not comprehensive, however they do provide an insight into the level of additional energy delivered to the vessel when all three options are combined.

7.4 WEATHER TRACKING & GREEN CORRIDORS

All wind propelled vessel have a weather station installed onboard which assists with the optimisation of the wind system performance in real time. Overlaying that with precise satellite weather forecasting data and LiDAR tracking for larger rig systems, there is an opportunity to both further improve the performance of the installed rigs, but also to inform designers on ways to improve design and providing extensive performance data to feed back into digital twin models and datasets available for improving the routing software.
This focus on preferable, optimised routes also aligns with the current development of 'Green Corridors'. There was significant movement at COP26 in Nov 2021 with the signing of the Clydebank declaration by 24 countries to establish green corridors to help facilitate the transition to zero-emissions shipping. While this declaration suggests a primary concern to establish the extensive and costly infrastructure to deliver new alternative fuels, the signatories did however pledge.
"In supporting the establishment of green corridors, signatories recognise that fully decarbonised fuels or propulsion technologies should have the capability to not add additional GHGs to the global system through their lifecycle, including production, transport or consumption. [28]
Thus, this identification of propulsion technologies fully aligns with the development of 'Green Corridors' that would take into account traditional sailing routes following the 'trade winds' and other prevailing wind routes. which can deliver substantial wind energy resources.

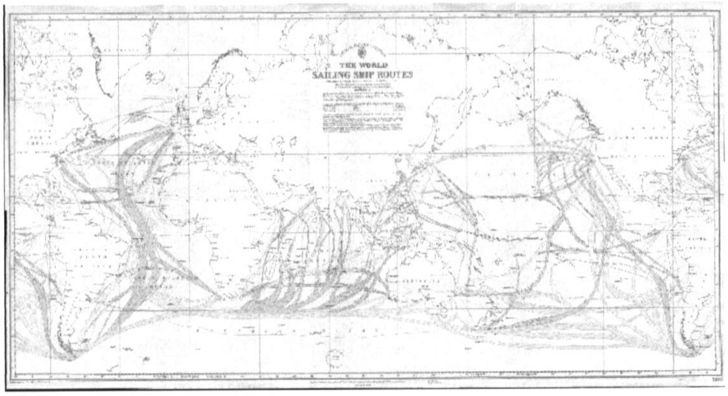

Picture 3. The Admiralty Chart 5308: World Sailing Ship Routes (Wikimedia commons)

These routes depicted in Picture 3 are very well known, predictable and backed with decades of weather data and thus can be once again used to facilitate low-emission corridors and integrated into this approach. [29] The driver for this to happen will be the need to meet ever tighter emission reduction target and will of course require operational profile adjustments, however in a carbon constrained and high fuel cost environment, this is a network of significant energy conveyers that can't be ignored.
The fact that this energy resource has been almost completely absent from the planning of existing green corridor development this year, clearly indicates the 'fuel-centric' approach rather than an 'emissions or energy- centric. [30] The latter approach has the potential to affectively speed up the move towards zero-emissions on these routes by lessening the fuel requirement for vessels and one which has the opportunity to unlock substantial savings in the near term that can be used to cover the additional investments required for the costly alternative fuel equipment and infrastructure. [31]

8. CONCLUSIONS

Just as commercial captains in the late 19th and early 20th Century wouldn't have dreamt of attempting ocean crossings without a wind-auxiliary system or at the very least an emergency sail canvas and rigging in case of steam engine failure, so this may well be so for the shipping of the future, where emission reduction, cost, and customer pressure to deploy readily available solutions are the driving factors shaping the fleet of the 2050s.

The new raft of wind propulsion technology deployments has, after a decade of development, trialling and demonstrating, started to take the first steps on a well-trodden pathway of technology scaling and diffusion and many of the perceived negatives or disadvantages associated with wind propulsion of the past have been reduced, become manageable or eliminated entirely.

The tweaking of the existing fleet through retrofits of wind-assist systems is already a trend that is underway, and it is one that will extend the carbon budget available to the industry and these WPT will not only pay for themselves but also release substantial funds needed to pay for further decarbonisation efforts through fuel savings. The move towards a transition is also underway, where there is a growth in new build primary wind vessels and optimised wind-assist designs along with wind-ready uptake which will increasingly be standardised. In a final step, transformation would include the development of a new paradigm in shipping, with primary wind designs being the standard for new, zero-emission vessels and an increasing number of those would be able to harvest excess wind energy to use a fuel when there are limited wind resources available to use directly.

While we have no crystal ball, we can state categorically that direct wind energy will always remain pure zero-emission, delivered for free directly to the point of use without the requirement of any infrastructure or storage thus making this a 'no regret' technology segment, or put quite simply a 'Win-Win-Wind' proposition.

7. REFERENCES

1. International Windship Association, Multi-stakeholder workshop report, June 2021

2. IMO MEPC 79 Inf. 21, 'Wind Propulsion' submitted by Finland, France, Saudi Arabia, Solomon Islands, Spain, Union of Comoros, RINA and the International Windship Association, October 2022

3. International Windship Association, Members Survey May 2022 (Confidential)

4. CE Delft, Study on the analysis of market potentials and market barriers for wind propulsion technologies for ships, Dagmar Nelissen (CE Delft) Michael Traut (Tyndall Centre) Jonathan Köhler (Fraunhofer ISI) Wengang Mao (Chalmers University) Jasper Faber (CE Delft) Saliha Ahdour (CE Delft), Nov 2016

5. MMMCZCS, Position Paper: Fuel Option Scenarios, Mærsk McKinney Møller Center for Zero Carbon Shipping, October 2021

6. Stopford, M, 'Stopford on Maersk and methanol' Interview with Dr Martin Stopford, Splash 24/7, 25 August 2021

7. OECD, The Impacts of Globalisation on International Maritime Transport Activity: Past trends and future perspectives, James J. Corbett and James Winebrake, Energy and Environmental Research Associates, the United States. Presented at the OECD/ITF Global Forum on Transport and Environment in a Globalising World 10-12 November 2008, Guadalajara, Mexico

8. Schinas, O, Sonechko, D, A pay-as-you-use business model for the greening of shipping, Cleaner Logistics and Supply Chain, Volume 4, July 2022

9. France Supply Chain, A New Step Forward: The selection of a Candidate by the Shippers to Deploy a Sail Container Door Service. 22 July 2022

10. MOL, "Wind Hunter Project" Zero Emission Ship Sailing by Wind and Hydrogen succeeded in the demonstration experiment by the Yacht "WINZ MARU", Mitsui O.S.K. Lines, 02 November 2021

11. IMO MEPC 79 Inf. 21, 'Wind Propulsion' submitted by Finland, France, Saudi Arabia, Solomon Islands, Spain, Union of Comoros, RINA and the International Windship Association, October 2022

12. CE Delft, Study on the analysis of market potentials and market barriers for wind propulsion technologies for ships, Dagmar Nelissen (CE Delft) Michael Traut (Tyndall Centre) Jonathan Köhler (Fraunhofer ISI) Wengang Mao (Chalmers University) Jasper Faber (CE Delft) Saliha Ahdour (CE Delft), Nov 2016

13. Frontier Economics, Reducing the Maritime Sector's Contribution to Climate Change and air Pollution: Economic Opportunities from Low and Zero Emission Shipping. A Report for the UK Department for Transport, Frontier Economics, July 2019

14. GIZ, Technical and Operational Options Catalog: Proposal for Technical and Operational Options to reduce Fuel Consumption and Emissions from Inter-Atoll-Transport and Inside-Lagoon-Transport, August 2019,

15. FAO., The State of World Fisheries and Aquaculture 2020. Sustainability in Action. Rome. 2020

16. IMO, Initial IMO Strategy on Reduction of GHG Emissions from Ships - Resolution MEPC.304(72) (Adopted on 13 April 2018)

17. Karslen, R, Papachristos, G, Rehmatulla, N - The diffusion of wind propulsion technologies in shipping: an agent-based model, 9th International Sustainability Transitions, Manchester, UK May 2018

18. International Windship Association, Multi-stakeholder workshop report, June 2021

19. WASP Project, Barriers & Overcoming Strategies for Accelerating the Uptake of WASP, December 2021

20. IMO MEPC.1/Circ.896, Guidance on Treatment of Innovative Energy Efficiency Technologies for Calculation and Verification of the Attained EEDI and EEXI, 14 December 2021

21. van der Kolk, N, Mason, J, Bonello, JM, Bordogna, G, Wind-Assist for Commercial Ships: A Techno-Economic Assessment, Wind-Assisted Ship Propulsion: SAIL ASSIST GROUP, 2019

22. International Council on Clean Transportation & Douglas Stolz, Cross Product Atmospheric LLC, Decarbonizing Bulk carriers with Hydrogen Fuel Cells and wind-Assisted Propulsion: A Modelled Case Study Analysis 2022

23. Paakkari, V, Wang, H, Stigler, C, Voyage Optimization with Wind Propulsion, RINA, International Conference on Computer Applications in Shipbuilding 13-15 September 2022

24. WASP Project, Speed trials for MV Copenhagen (Rotor Sail), MV Frisian Sea (Suction wings) & MV Annika Braren (Rotor sail), Project Deliverables, 2022

25. Olsson, T, Carlsson, J, Life cycle modelling of a wind powered car carrier; An assessment of cost and greenhouse gas emissions, Master's thesis in Maritime Management, 2020

26. de Ricqlès, J, Pros and cons of sail-powered container ships, Upply Market Insights, September 29, 2022

27. Mason, J, Quantifying voyage optimisation with wind-assisted ship propulsion: a new climate mitigation strategy for shipping, James C Mason PhD, University of Manchester, 2021

28. Clydebank Declaration, COP26 Glasgow, UK Government Publications, November 2021

29. Rodrigue, J-P, The Geography of Transport Systems, Routledge, 2020

30. Global Maritime Forum, Annual Progress Report on Green Shipping Corridors, November 2022

31. Allwright, G, Could wind propulsion fund the full decarbonisation of shipping? Splash 24/7, 15 March 2021

8. AUTHORS BIOGRAPHY

Gavin Allwright holds the current position of Secretary General at the International Windship Association. He is responsible for the overall running of the association, along with representation at policy fora such as the IMO, EU and at national levels. His previous experience includes work with a small zero-emissions vessel development team and research and lecturing in sustainability and the application of zero-emissions shipping in LDC and SIDS regions. He holds a Masters degree in Sustainable Development, specialising in small scale sustainable shipping and logistics in developing countries.

THE USE OF MARINE NUCLEAR POWER FOR THE PROVISION OF SYNTHETIC FUELS 'E-FUELS'

G Gennaro and **R Megginson**, Core Power, UK

SUMMARY

Shipping is responsible for approximately 3% of global emissions and consumes about 380 million tons of HFO equivalent per year. E-Fuels are considered an important pathway to decarbonization, however for them to replace fossil fuels their production must be affordable, have very low emissions and externalities from well to wake, and scalable. Modelling the production of e-Fuels indicates that Floating Nuclear Power Plants coupled with Floating Production and Storage Units, are the best option to fulfil these criteria. In particular nuclear power has been historically characterised by very high-capacity factor, low OPEX, and very low externalities, and by combining advanced modular reactors with offshore installations taking advantage of modular and serial shipyard construction a dramatic reduction in CAPEX would be achieved. These facilities could be deployed worldwide near major shipping hubs, i.e. near the point of bunkering, drastically reducing the cost and the complexity of the supply chain.

1. INTRODUCTION

Shipping is responsible for approximately 3% of global emissions. The IMO has set an ambitious decarbonisation goal of a 50% reduction in emissions by 2050 and is expected to further strengthen this goal to push towards full decarbonisation. The industry is currently seeking means of achieving these goals. E-fuels such as ammonia, methanol and synthetic fossil fuels are a potential area of interest for the transition to zero-emission shipping. By producing these fuels from renewable nitrogen, hydrogen, and CO_2 it means that they have the potential to be carbon neutral offering a path towards decarbonisation for shipping as well as providing an alternative fuel for the aviation and manufacturing industries [1].

However, the maritime industry cannot think of the requirement for synthetic fuels in isolation, it will be part of a wider energy system, therefore the production methods for these fuels will have to match not just maritime but also aviation and potential competition from steel and cement manufacturers. Aviation uses approximately 329 million tonnes of jet fuel per annum [2] compared to shipping's 340 million tonnes of HFO equivalent [3]. A comparison of the equivalent volumes of these fuels needed if replaced by e-fuels is summarised in Table 1.

Table 1: Comparison of volumes of potential e-fuels needed to meet current energy requirements of the maritime and aviation sector

	Traditional Fossil Fuels	Ammonia	Methanol	E-Methane
Maritime	340 million tonnes (HFO)	783 million tonnes	707 million tonnes	267 million tonnes
Aviation	329 million tonnes (Jet Fuel)	506 million tonnes	503 million tonnes	281 million tonnes

Table 1 clearly shows that ammonia and methanol will require significantly more fuel compared to traditional fuels due to their lower energy density. Even when considering E-Methane which lowers the mass of fuel needed due to higher energy density, the amount of green hydrogen will need an order-of-magnitude change in the current production levels. These very large quantities required mean that the scalability of any potential e-fuel / production process is a key parameter when considering the economic feasibility and uptake for the global energy system.

Therefore, for electro fuels to be significant, their production method must fulfil three criteria:

- Affordability. For e-Fuels to be used by the shipping, aviation and land-based industries, they must be competitive with existing fuels otherwise dirtier but affordable fuels will be used. Feedstock cost, capacity factor, availability and dispatchability of the necessary power generation, as well as the logistic infrastructure, are all key factors in this calculation.

- Minimal Total Life Cycle Externalities. E-Fuels should not only reduce direct greenhouse gas emissions, but also indirect emissions and total life cycle externalities. A well-to-wake approach is needed to avoid moving emissions and externalities upstream rather than reducing them.

- Scalability. The production of e-Fuels needs to scale to the size of the demand. This requires understanding the amount of natural resources, raw materials and land needed to scale production to meet all the demands of the industry.

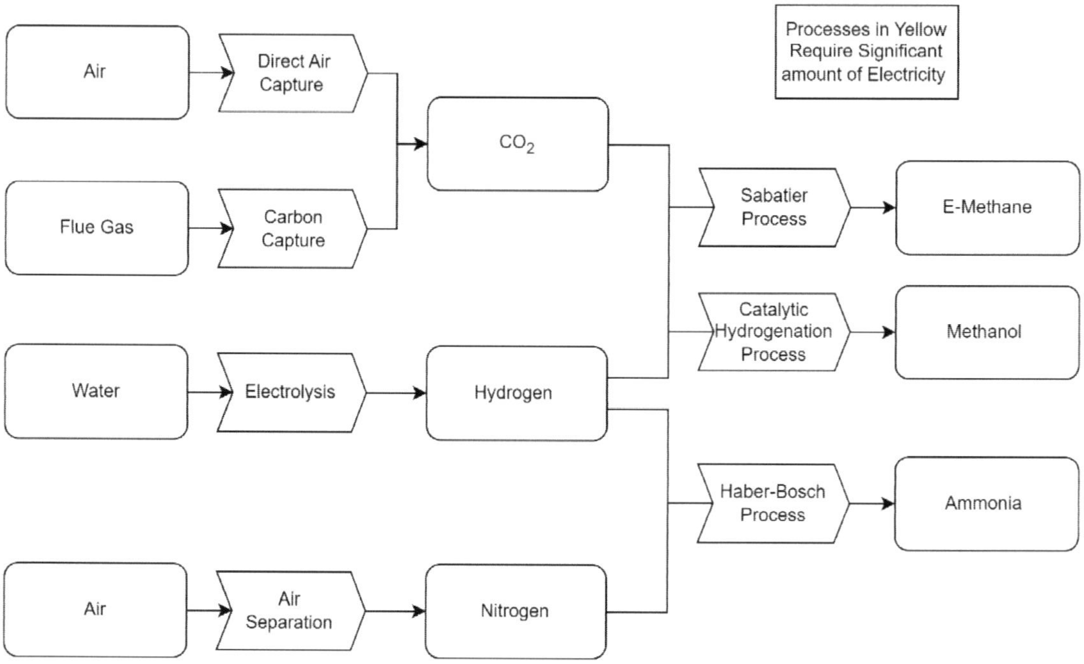

Figure 1: Summary of pathways for e-fuels for shipping.

While E-fuels have seen a sustained and growing interest as a potential replacement fuel for international shipping there has also been an increasing awareness of the potential problems [4]. The best way to provide the capacity of electricity generation required for this process is developments in the next generation of floating electrical nuclear power. This can provide the reliable and scalable source of power needed to meet the demand for next-generation fuels.

2. THE E-FUEL PRODUCTION PROCESS

The key step in the e-fuel process is the production of the hydrogen used in all these synthetic fuels. For these fuels to have the low externalities that will be demanded by environmental regulations it is important that this hydrogen comes from a source without large downstream emissions such as steam methane reforming. The need for this hydrogen is therefore best met by green hydrogen production.

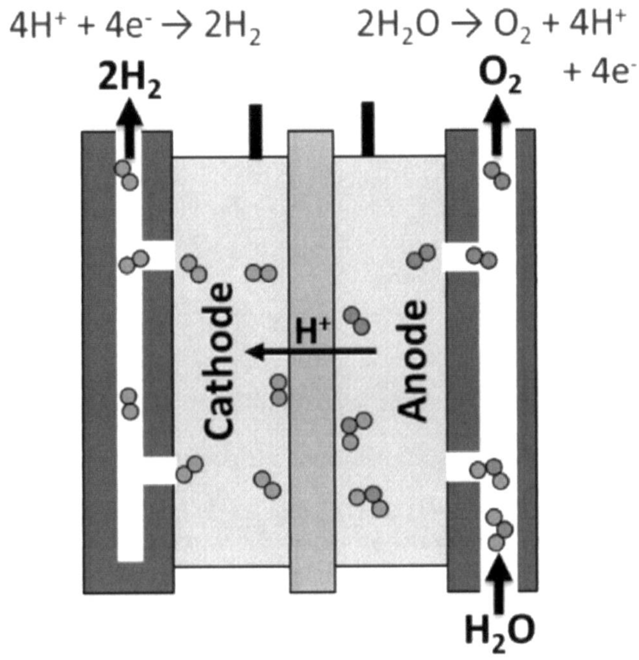

Figure 2: A schematic of a Proton Exchange Membrane Electrolyser

Green hydrogen is where electricity is used to split water molecules into their constituent hydrogen and oxygen components. The process has been in operation commercially for almost a century, however, there have been significant increases in investment as hydrogen has been seen as an alternative energy carrier to fossil fuels in a fully decarbonised energy system. For this hydrogen to be considered truly green however it must be derived from a green energy source and not merely have moved carbon emissions upstream.

Once the hydrogen is produced it can be combined with Nitrogen or CO_2 to produce the final e-fuel

2.1 AMMONIA

Ammonia has seen widespread industrial manufacture for over a century as an important component of the global fertiliser industry. For marine production, the process will be able to use abundant air and seawater. Hydrogen from the electrolysis of desalinated seawater will be combined with atmospheric nitrogen in a Haber Bosch process to produce ammonia.

$$N_2 + 3H_2 \rightarrow 2NH_3$$

Equation 1 Ammonia Production

One of the most important characteristics of this process is that all of the hydrogen produced is used in the final fuel increasing the overall energy efficiency of the system.

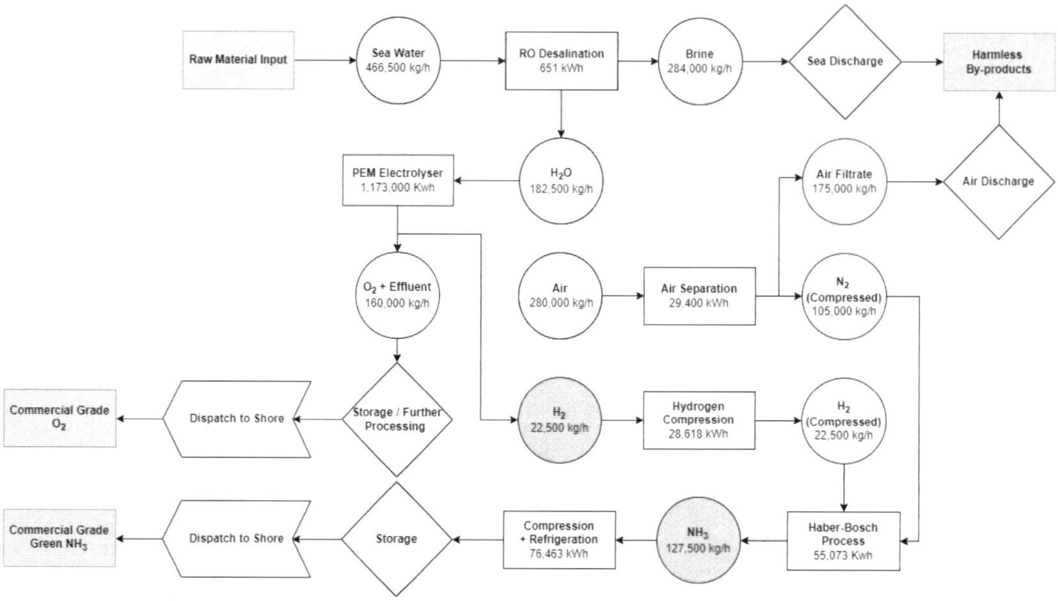

Figure 3: Flow diagram for an ammonia production of 1 million tpa of ammonia.

Ammonia has received significant interest as a marine fuel as well as a potential alternative fuel for other hard-to-electrify sectors. Ammonia has many attractive qualities as an alternate fuel. Firstly, it is already widely manufactured and transported, as part of the global fertiliser industry. Ammonia also has the benefit of being zero carbon emission at the point of ignition, though does produce some NO and NO_x which would have to be managed with changes in scrubber technology. There are however concerns about the safety management of large amounts of ammonia [2].[ii] The world's first ammonia fuelled vessel was Kriti Future a 156,000 DWT Tanker delivered in 2022 by New Times Shipbuilding to Greek shipowner Avin International

2.2 METHANOL

Currently, methanol is manufactured by reforming natural gas, however, with the global drive for decarbonisation, there has been increasing interest in manufacturing methanol from CO_2. The first commercial plant to use this method, the George Olah in Iceland, was commissioned in 2012 and is now capable of producing 4000 tonnes of methanol per year [5].

$$CO_2 + 3H_2 \rightarrow CH_3OH + H_2O$$

Equation 2 Methanol Production

For this process to operate effectively a reliable and sufficiently large source of CO_2 will need to be sourced. Current discussions have discussed the use of CO_2 captured from burning organic material (BECCS) or captured from fossil fuel use. There have been suggestions that these fuels could be made from CO_2 removed from the atmosphere however the high power and space requirements of this process make it economically unattractive [6].

Methanol has the benefits of being both denser and more energy dense than ammonia. The major problem though will be sourcing sufficient quantities of green CO_2 from either biological sources or from carbon capture. There is also a loss of overall energy efficiency as $1/3^{rd}$ of the hydrogen needed for the reaction is wasted in the formation of water as part of the reaction. The development of methanol for shipping is being driven forward by the A. P. Moller Maersk group, who have put significant investment into the development of the infrastructure for methanol production [6].

2.3 E-METHANE

Another potential option for alternative fuels is the production of E-Methane. E-Methane's major benefit is that Methane, in the form of LNG, is already being used widely in the industry for propulsion. LNG had been seen as a more cost-effective option compared to traditional fuel while also lowering emissions, however with recent increases in gas prices this economic advantage has decreased. Of all e-fuels E-Methane currently has the largest number of possible vessels capable of using it as a fuel with over 800 vessels expected to be capable of using LNG by 2028 [7].
Synthetic Methane can be made through the following process known as the Sabatier Reaction

$$CO_2 + 4H_2 \rightarrow CH_4 + 2H_2O$$

Equation 3. The Sabatier Process

The reaction is normally carried out at high temperature, 300-400 °C, and pressure, 3 MPa, normally in the presence of a ruthenium catalyst. While this a one step process it does suffer a loss of overall efficiency as ½ of the hydrogen required for the process is wasted in the production of water, with the loss being stronger than in the methanol production process [8].

E-methane also suffers from the same problems as E-Methanol in regard to procuring sufficient renewable CO_2 as a feedstock for the process.

Within the maritime industry CMA CGM has joined the Jupiter 100 project with various French companies to produce e-methane from green hydrogen from renewable electricity and CO_2 from industrial Carbon Capture [9].

3. Why Nuclear

Green hydrogen from nuclear power is the only technology choice capable of meeting the three criteria for e-fuels. Nuclear power is already the safest and most reliable form of electricity generation in operation globally. However traditional nuclear has suffered from high build costs and long deployment times.

The advent of next generation reactors however such as the Molten Chloride Fast Reactor (MCFR) offer the opportunity to move to shipyard construction of nuclear reactors reducing both cost and time to deployment. This process does not merely displace emissions upstream to the point of electricity provision but minimise externalities throughout the production lifetime.

The MCFR is an advanced nuclear reactor powered by liquid molten fuel salts rather than solid fuel elements assembled in rods like those in conventional pressurised water reactors (PWR). The MCFR operates at ambient pressure only and runs at a high temperature, between 500°C and 700°C.

The fuel salt contains the nuclear materials, producing both the heat and the coolant which transports the heat to the conversion system producing electric power. Because the MCFR operates at ambient pressure there is no force to expel radio toxins into the environment in the event of an accident. This allows for a very small emergency preparedness zone (EPZ), the area for which contingency plans must be put in place in case of the release of radioactive material. The EPZ of facilities would not extend beyond the boundary of the facilities hull simplifying the licencing and deployment process If the reactor stops and the fuel salt is cooled it would solidify and be entombed in the reactor. This means no fuel can escape. This passive safety feature of the MCFR is highly desirable for the marine environment.

Nuclear power also offers the highest capacity factors and footprint to power ratio of all green electricity generation technologies summarised in table 2. This is vitally important in meeting the scalability challenge of e-fuel manufacture.

Table 2: Comparison of capacity factors [10] and land use for green electricity generation technologies [11] and CO_{2eq} [12]

Generation Method	Capacity Factor / %	Land Use / m^2/MWh	CO_{2eq}/GWh
Nuclear	93	0.3	3
Hydro	41	33	34
Offshore Wind	35	128	4
Solar	24	19	5

Existing reactors already have the highest capacity factors of any generation method and the advances in the next generation in nuclear will allow reduction in cost and deployment time while achieving higher standards of safety and security.

4. THE FLOATING NUCLEAR REFINERY

CORE POWER are proposing the deployment of a number of offshore structures to be able to manufacture these fuels. The CORE POWER concept envisages a floating MSR power station producing 1.2 GW of electricity providing power to a separate fuel production structure/barge. These fuels can then either be transported to storage tanks or directly to vessels.

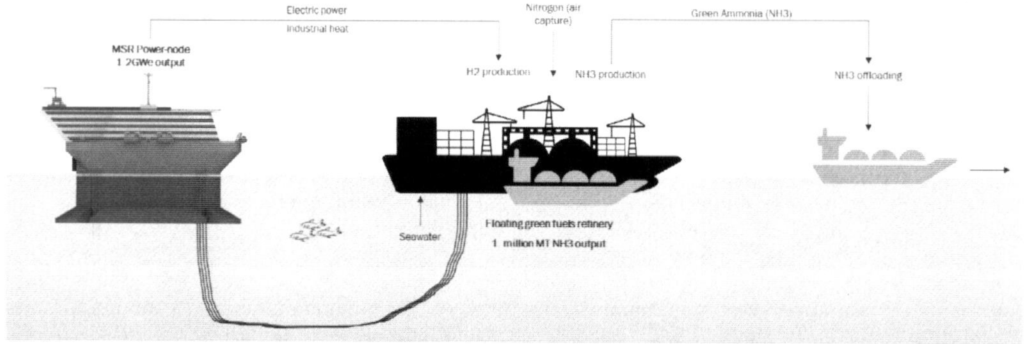

Figure 4: Schematic Diagram of proposed CORE POWER facility for the production of e-Ammonia

Modelling this system based on data from currently available industrial plant suggests that using 1.2 GW of Electric power could produce between 100,000-450,000 tonnes of HFO equivalent of fuel.

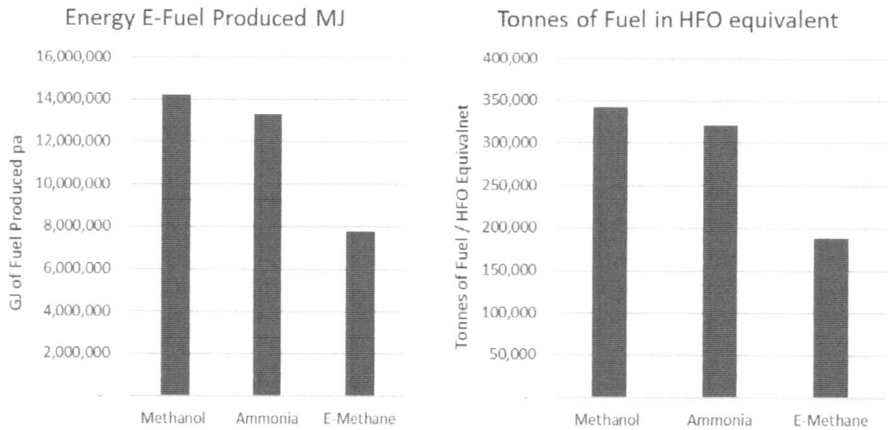

Figure 5: Comparison of Tonnes of E-fuel and Energy content of E-fuel produced per annum using marine nuclear capable of producing 1.2 GW_e of power assuming a 95% capacity factor.

Figure 5 shows the amount of fuel that it is capable to manufacture using the 1.2MW_e output of a CORE POWER floating MCFR facility. Methanol is the most energy efficient fuel that can be produced, equivalent to 340,000 tonnes of HFO. While wasting some hydrogen in the production of water its higher energy density offsets this to make it the best option. Ammonia is the second-best option producing the equivalent of 320,000 tonnes of HFO. E-methane is the least efficient production product, producing the equivalent of 190,000 tonnes of HFO, significant lower due to it high amount of wasted hydrogen as part of the production process. One of these nuclear facilities would be capable of producing the equivalent of 0.1% of global shipping fuel a significant step towards decarbonisation.

By operating these facilities offshore, they would have a number of advantages compared to land-based systems. These facilities will be able to be built in shipyards decreasing the time and cost of construction as well as giving flexibility of deployment. These facilities could be envisaged to be deployed in major shipping and aviation hubs where fuels are required such as Rotterdam, Houston, and Singapore.

5. ECONOMICS OF MARINE E-FUEL PRODUCTION

Using the proposed levelised cost of electricity for floating advanced nuclear of 45 $/MWh it is possible to estimate the potential costs of production for Ammonia, Methanol and E-Methane using the standard values summarised in Table 3.

Table 3: Economic Values used in price Calculations

Electricity Price Floating Nuclear	45 \$/MWh
Nitrogen Price	100 \$/Tonne
CO_2 Price (BECCS)	150 \$/Tonne
Electrolyser	PEM Electrolysers
Ammonia Energy Density	18.6 MJ/Kg
Methanol Energy density	19.5 MJ/Kg
E-Methane Energy Density	52.75 MJ/Kg
Haber-Bosch Capex Cost	300 \$/Tonne
Sabatier Capex Cost	115 \$/Tonne
Catalytic hydrogenation CAPEX Cost	500 \$/Tonne

The results of this modelling are presented in figure 6 comparing the cost for both 1 tonne and 1 MJ of fuel. The most cost-effective option is found to be methanol due to its higher density and energy density mean that it can be produced at about 630 \$/Tonne.

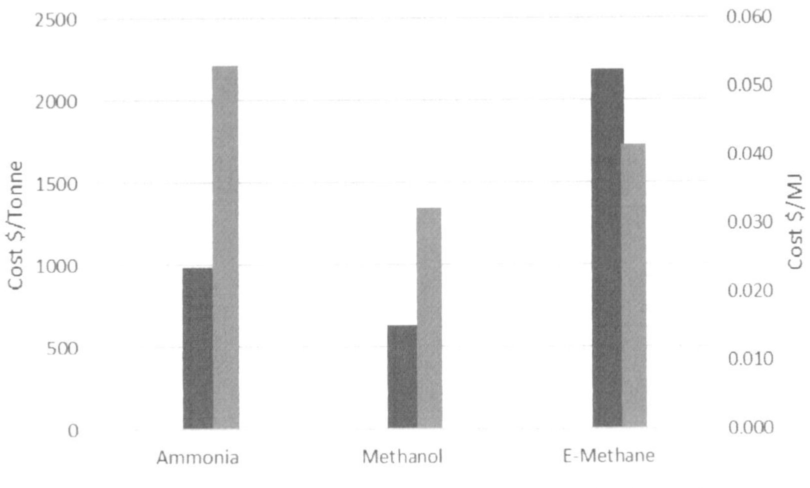

Figure 6: Cost Comparison of three major candidates for E-fuels made with marine nuclear power

Marine nuclear power offers the most competitive option for the production of e-fuels when compared to the levelised cost of electricity for intermittent renewable with the storage needed to make them reliable of 117 \$/MWh [13] or the cost of Dutch grid electricity for industry of 312 \$/MWh.[14] Both these options would vastly increase the cost to produce these fuels as is shown in Table 4.

Table 4: Comparison of the production cost of alternative fuels, in $/Tonne, using different electricity prices

Fuel	Advanced Marine Nuclear	Renewable and Storage	Current Netherland Grid
Ammonia	990	1,870	4,245
Methanol	630	1,530	4,160
E-Methane	2,190	4,585	11,070

The above table shows why marinised nuclear power is the best possible option for the provision of e-fuels it is not only the most reliable option but also the only one that will be able to offer the fuels that will be acceptable to the international maritime industry. The total system cost of grid electricity and even specifically co-sited renewable paired with storage will not be able to produce the fuels at the costs needed.

6. CONCLUSIONS

The coupling of Floating Nuclear Power Plants with Floating Production and Storage Units is possibly the dominant option for the large-scale production of electro fuels, not only for marine uses but also for aviation and land-based industry. This conclusion is based on the fact that this is, by far, the only technology capable of meeting the three criteria needed for the effective and large-scale deployment of synthetic e-fuels.

- Affordability. Marine nuclear is able to offer electricity at a lower price than grid-based electricity or renewables and storage. With expected developments in electrolyser technology, it will be possible to compete with traditional fuels by 2050.

- Minimal Total Life Cycle Externalities. Marine nuclear has no hidden upstream emissions and will offer low externality electricity production throughout its lifespan.

- Scalability. By utilising shipyard construction, it will be possible to increase the deployment times of these facilities when compared to land-based deployments. Furthermore, these facilities will use much less space than an equivalent renewables-based project.

CORE POWER envisages a series of such facilities deployed close to major shipping hubs and producing efuels at the point of bunkering, facilitating the creation of green corridors, one of the leading proposals for the development of a green shipping ecosystem, and further reducing the cost and complexity of the efuels supply chain.

9. REFERENCES

1. M. Yugo and A. Soler, "A look into the role of e-fuels in the transport system in Europe (2030–2050)", Concawe Review, vol. 28, pp.4, 2019
2. EIA. Jet Fuel Consumption, World, Annual [Online] Available https://www.eia.gov/opendata/v1/qb.php?category=2135044&sdid=INTL.63-2-WORL-MT.A
3. IMO MEPC, "Energy Efficiency of Ships-Report of fuel consumption data submitted to the IMO Ship Fuel Oil Consumption Database in GISIS", 2021
4. Together in Safety, "Future Fuels Risk Assessment", 2022
5. IRENA, "Renewable Methanol", 2021
6. MAERSK (2022 March. 10), A.P. Moller - Maersk engages in strategic partnerships across the globe to scale green methanol production by 2025. Available https://www.maersk.com/news/articles/2022/03/10/maersk-engages-in-strategic-partnerships-to-scale-green-methanol-production
7. Riviera, Record-setting 2021 to swell LNG-fuelled fleet to over 800, https://www.rivieramm.com/news-content-hub/news-content-hub/record-setting-2021-to-swell-lng-fuelled-fleet-to-over-800-69308
8. Stefan Rönsch et al, Review on methanation – From fundamentals to current projects, Fuel, Vol 1666 , 2016
9. MA CGM to run its ships on e-methane, https://lafrench-fab.com/news/cma-cgm-to-run-its-ships-on-e-methane/
10. Eia, Electric Power Monthly, October 2021
11. Our world in Data, How does the land use of different electricity sources compare?, https://ourworldindata.org/land-use-per-energy-source
12. Our World in Data, What are the safest and cleanest sources of energy?, https://ourworldindata.org/safest-sources-of-energy
13. Lazard (2021 Oct. 8), "Levelized Cost Of Energy, Levelized Cost Of Storage, and Levelized Cost Of Hydrogen" Available; https://www.lazard.com/perspective/levelized-cost-of-energy-levelized-cost-of-storage-and-levelized-cost-of-hydrogen/
14. Global Petrol Prices (2022),Electricity Prices, Available https://www.globalpetrolprices.com/electricity_prices/
15. A. Bielecki et al, "The externalities of energy production in the context of development of clean energy generation", Environmental Science and Pollution Research, 27, 11506-11530

8. AUTHORS BIOGRAPHY

Giulio Gennaro is the Chief Technical Officer of CORE POWER. He is a Chartered mechanical engineer from the University of Genova, Italy, and he holds and EMBA from SP Jain School of Global Management, Singapore. He is an expert in ship propulsion, machinery, asset integrity and all related to failure investigations of machinery and mechanical components. He is also a mentor for Pier 71, the joint incubator and accelerator by the Maritime and Port Authority of Singapore and the National University of Singapore.

Rory Megginson is the Director of Analytic of CORE POWER. He Is a former academic researcher with a passion for decarbonisation, who joined CORE POWER after completing a PhD in Chemistry at the University of St Andrews. Rory heads up the Analyst team at CORE POWER and his main area of focus is the creation of green fuels using marinised nuclear power.

THE GUTTA-VISIR SYSTEM FOR OPERATIONAL CO_2 EMISSION SAVINGS FROM FERRIES

G Mannarini, L Carelli, F Viola, M Hoxhaj, M Scuro, S Kyal, R Lecci, G Coppini, Centro EuroMediterraneo sui Cambiamenti Climatici (CMCC), Italy

SUMMARY

GUTTA-VISIR is a service for operationally providing least-CO_2 ferry routes. It combines numerical meteo-oceanographic forecasts with both the ship route optimisation model VISIR and a dedicated environment for operational production. The regions of interest are the Adriatic and Northern Ionian Seas. The optimal routes are computed at the CMCC high-performance computing facility by means of the VISIR model and then made available through a dedicated web application (www.gutta-visir.eu). Both the CO_2 and the Carbon Intensity Indicator (CII) savings resulting from route optimisation are provided to the end users. GUTTA-VISIR has been in continuous operation since November 2021 and 39,600 among the routes published since were statistically assessed.
It is found that the maximum CO_2 savings for a given route are related to the average sea state, that the CO_2 savings on any given route are nearly exponentially distributed, and that the largest mean savings are obtained for routes in the southern part of the basin. While the average savings for these short-sea routes in a relatively calm basin are around one percent, two-digit savings relative CO_2 emissions can be obtained on specific dates and routes. Furthermore, it is seen that the route optimisation process can account for about one tenth of the CII reductions required in 2023 by the International Maritime Organization.
These results provide a first objective assessment of the potential of ship routing for decarbonisation of short-sea shipping.

NOMENCLATURE

α	angle of attack of waves (relative true angle between waves and course)
CII	carbon intensity indicator (various units)
CMS	Copernicus Marine Service
CO_2	carbon dioxide emissions (t)
χ	engine fractional load (%)
$-\Delta CO_2$	CO_2 relative savings (with respect to the geodetic route, %)
EEA	European Economic Area
GHG	greenhouse gas(es)
GT	gross tonne
GUTTA	savinG fUel and emissions from mariTime Transport in the Adriatic region
HPC	high performance computing
Hs	significant wave height (m)
SOG	speed over ground (kts)
STW	speed through water (kts)
VISIR	discoVerIng Safe and effIcient Routes
w	surface ocean current intensity (kts)

1. INTRODUCTION

We live at the time of a climate crisis. The United Nations (UN) through several agencies have reviewed the scientific foundations (IPCC), the impacts (WMO, UNEP), and the policy instruments (UNFCCC) to address such a crisis. The general understanding is that humanity is lagging in taking action to remove the cause of this crisis. Namely: a deep alteration of the global carbon cycle, due to the direct emission of large amounts of greenhouse gases (GHG) through combustion of fossil fuels and cement production, as well as to the reduction of the GHG removal processes from the atmosphere, due to changes in land use, land cover, and forestry [5]. The resulting enhancement of the natural greenhouse effect has led to global warming, with consequences in terms of environmental degradation, natural disasters, weather extremes, food and water insecurity, economic disruption, conflicts, and climate migrations. To change course, together with nature-based solutions, support to scalable decarbonisation technologies was recently emphasised by the UN[1]. This must hold for the sector of shipping too. Shipping as such was not explicitly included in the resolutions of the Paris

[1] https://www.un.org/en/un75/climate-crisis-race-we-can-win

Agreement[2]. However, the UN pushed the maritime sector to deliver a strategy for contributing to decarbonisation. This was agreed upon by the International Maritime Organization (IMO) in 2018 [6].
IMO also approved the first concrete measures to be applied in the short term. Specific metrics, called Carbon Intensity Indicators (CII) have been introduced. CIIs all take the form of a CO_2 amount divided by some representation of transport work. The baseline for the mandated CII reduction is the year 2019, and the reduction rates increase every year till 2026. Subsequent reductions have still to be agreed upon[3]. While several Member States noted that the IMO Initial Strategy will be insufficient to deliver a contribution in line with the Paris Agreement, a revision of the IMO initial strategy is due by 2023 and should enter in force by 2027 [7].
In the meantime, the European Union (EU) has made independent steps to accelerate maritime decarbonisation. The CO_2 emission data of all vessels above 5,000 GT calling at EEA harbours have been published since 2018 [11]. This database shall be used for determining the allowances due by the ship companies, once the sector will be part of the EU-Emissions Trading System. This was proposed in the frame of the fitfor55 legislative package[4]. At the current price[5] and CO_2 emission level, the total value of allowances for shipping in the EEA can be estimated to be in the order of ten billion Euro per year.
During the last UNFCCC conference in Glasgow, several UN Member States agreed to establish at least six green corridors by the middle of this decade ("Clydebank declaration" at COP26[6]). Fully decarbonised fuels or propulsion technologies should be available at ports of a green corridor.
All these measures and proposals will necessarily make shipping more expensive than today. Thus, saving both fuel consumption and GHG emissions will be increasingly important. This is one of the reasons why voyage optimization or ship routing can play a role in facilitating both the energy transition and the economic viability of a low-carbon shipping industry. Furthermore, voyage optimization can apply to both newbuilds and retrofits of existing vessels. However, it is often difficult to estimate to what extent it could impact a decarbonisation pathway. Indeed, academic literature usually focuses on case studies, while data from commercial routing systems are not available for assessment. To close this knowledge gap, open access to the source code of a routing model would be extremely beneficial. However, to go beyond the level of a case study, a model should be operationally executed.
This is exactly where this paper would like to contribute. CMCC has been developing the VISIR open-source model[7] for weather routing for the past ten years. During the Italy-Croatia Interreg project GUTTA[8] the VISIR model was further developed and used for operational production of least-CO_2 routes for ferries in the Adriatic Sea. This basin connects Italy to Slovenia, Croatia, Montenegro, Albania, and Greece. Its maximum width of about 110 nautical miles in direction normal to its elongation axis makes overnight ferry crossings possible. Therefore, the scientific and practical question arises, if even short sea shipping could appreciably benefit from ship route optimization. This question can be answered through the GUTTA-VISIR operational service. It has been operational for about one year now, and the data produced were systematically assessed, as documented in the remainder of this paper.
In Sect.2 the materials and methods used, i.e. the VISIR model (Sect.2.1) and the GUTTA-VISIR system (Sect.2.2) are presented. The results, in Sect.3, are subdivided into a short review of the GUTTA-VISIR web application (Sect.3.1) and an analysis of data collected, with respect to both the CO_2 and CII savings (Sect.3.2). Our conclusions are drawn in Sect.4.

2. MATERIALS AND METHODS

This section presents the key materials and methods developed for producing the results given later in Sect.3. The materials and methods consist of the VISIR-2 ship routing model, documented in Sect.2.1, and the GUTTA-VISIR operational service, described in Sect.2.2.

2.1 THE VISIR-2 SHIP ROUTING MODEL

VISIR-2 is a python software meant for computing optimal ship routes in presence of dynamic marine weather fields. It was documented in [8]. VISIR-2 was inspired and validated vs. its predecessor; the matlab-coded VISIR-1 model [9]. The main model components of VISIR-2 are a vessel response model, a processor of environmental fields, a path planning component, and a suite of visualisation routines. They are briefly reviewed in the subsequent paragraphs.

---

2 https://unfccc.int/process-and-meetings/the-paris-agreement/the-paris-agreement

3 https://www.dnv.com/news/imo-update-marine-environment-protection-committee-mepc-76-203128

4 https://climate.ec.europa.eu/eu-action/european-green-deal/delivering-european-green-deal/increasing-ambition-eu-emissions-trading_en

5 https://tradingeconomics.com/commodity/carbon

6 https://www.gov.uk/government/publications/cop-26-clydebank-declaration-for-green-shipping-corridors/cop-26-clydebank-declaration-for-green-shipping-corridors

7 www.visir-model.net

8 https://bit.ly/guttaproject

2.1 (a) Vessel response model

The vessel speed through water (STW) of a ferry was estimated from the outcome of a ship simulator installed at the University of Zadar (a Partner in the Italy-Croatia GUTTA project which also funded the development of GUTTA-VISIR). The hardware consisted of the Wärtsilä-Transas Marine NTPro 5000 Navigation Simulator coupled to the Wärtsilä ERS-LCHS 5000 TechSim engine-room simulator. The vessel was a Ro-Pax ferry with a twin, four-stroke engine (MAN Diesel 32/40 Twin Medium Speed) with controllable pitch propeller. Some vessel characteristics are provided in Tab.1. while for more technical details the reader can refer to [13].

Table 1: Parameters of the modelled ferry.

Name	Symbol	Value	Units
Length at waterline	Lwl	114	m
Hull coefficient	Cb	0.54	-
Deadweight	DWT	4,050	t
Gross tonnage	GT	14,000	t
Main engine power	Pme	4,000	kW
Main engine rated speed	n	750	rpm
Service speed	Vs	19	kts

The simulator was operated in the absence of sea currents, as they were added later as described in Sect. 2.1.c. Regular waves of significant height *Hs* and with a wave dispersion typical for the central Adriatic Sea were then generated, with variable angles of attack α with respect to the ferry. The engine load factor χ varied between 70 and 100%. The STW was a direct output of the simulator, while the CO_2 emission rate was estimated from the product of the specific fuel consumption and the power delivered at the propeller shaft, times a convenient emission factor for the actual fuel type used. The multi-dimensional response of the ferry (as a function of Hs, χ, and α) was published and commented on in [8].

2.1 (b) Environmental fields

As VISIR-2 is a graph-based model [8], it was necessary to define the graph edge weights to be later used by the path planner (Sect.2.1.c). To this end, the environmental fields from data-assimilative ocean models must be represented on graph edges. Also, for correctly capturing the field dynamics, they are sampled at a time resolution higher than in the original numerical products. This requires an interpolation, both in space and time. For the present application, forecast fields from the Copernicus Marine Service[9] (CMS) are used,

- The MEDSEA_ANALYSISFORECAST_WAV_006_017 sea state product (VHM0, VMDR variables),
- The MEDSEA_ANALYSISFORECAST_PHY_006_013 sea circulation product (uo, vo variables).

They both have a spatial resolution of 1/24 degree (2.5 miles in the meridional direction) and an hourly resolution (hourly instantaneous fields for waves, hourly mean values for currents). These products are computed from numerical models making use of data assimilation. Satellite sea wave height data are assimilated for the former and sea level, temperature-salinity profiles, sea surface temperature for the latter. The graph grid where the fields were remapped had a resolution of 1/12 degree, while the temporal resolution was 30 minutes.§

2.1 (c) Path planning component

The vessel's speed over ground (SOG) is formed from vector composition of the STW and the sea surface current velocity [9]. The STW is obtained from the vessel response (Sect.2.1.a) in correspondence of the actual significant wave height and direction from the dynamic environmental fields (Sect.2.1.b). Then, the SOG is used to compute the time for sailing across any graph edge, at each point in time. A path finding algorithm makes use of this information to extract the optimal routes. The algorithm used in VISIR-2 belongs to the family of graph-search methods [3]. Specifically, it is a modified version of the time-honoured Dijkstra's algorithm [4]. The algorithm was first generalised to deal with dynamic edge

[9] https://marine.copernicus.eu/

weights, proving its outcome vs. a semi-analytical benchmark [9]. Then, it was extended to also compute least-CO_2 paths [8]. As a matter of fact, three variants of this algorithm are now available in VISIR-2: a static one, to compute least-distance routes making use of just bathymetry and shoreline as constraints; two dynamic ones, to compute least-time and least-CO_2 routes.

2.1 (d) Visualisation

A typical graphical output from VISIR-2 is shown in Fig.1. It includes both optimal routes and environmental fields. The departure location is always portrayed as a star. The three optimal routes are displayed with different colours. The fields are displayed in grey tones. To provide dynamic information via a static picture, the fields are rendered with concentric shells originating at the departure location. The shape of these shells is defined by isochrones. These are lines joining all sea locations which can be reached from the origin upon sailing for a given amount of time. This way, at any route waypoint, the field (either significant wave height *Hs* or current magnitude *w*) is portrayed at the time the vessel is supposed to be at that location.

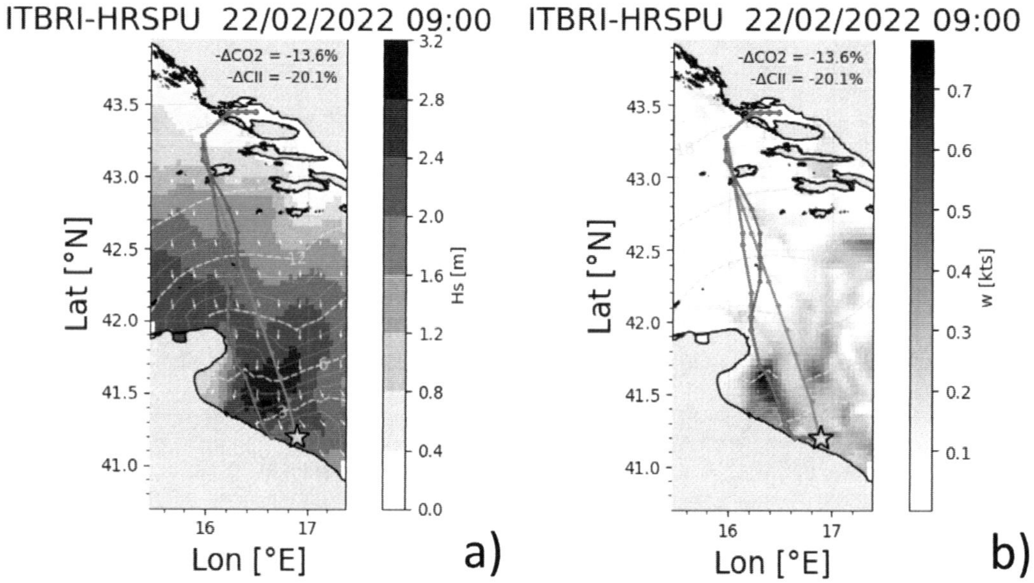

Figure 1: Optimal routes (blue/red/green, solid lines for distance/time/CO_2 respectively) and isochrones (golden, dashed lines) on top the significant wave height field (shaded grey tones). Between each pair of isochrones, the field is displayed with hourly resolution. CO_2 and CII savings of the green with respect to the blue route are given in the top right.

2.2 THE GUTTA-VISIR OPERATIONAL SERVICE

The GUTTA-VISIR[10] system was conceived to operationally compute and publish optimal ferry routes, based on marine forecasts. The service was developed in the frame of the Italy-Croatia Interreg project GUTTA[11] and was powered by the VISIR-2 model. Through an operational chain on a high-performance computing (HPC) facility and a dedicated web-application, GUTTA-VISIR daily provides an end-user with both graphical and numerical products. The access to the service is open and free. Fig.2 provides an overview of the system. The data flow originates at the daily (or bi-daily for waves) feed of CMS products for the Mediterranean Sea and ends on the user's device. Its back- and frontend components are shortly described in the following two paragraphs.

[10] www.gutta-visir.eu
[11] https://bit.ly/guttaproject

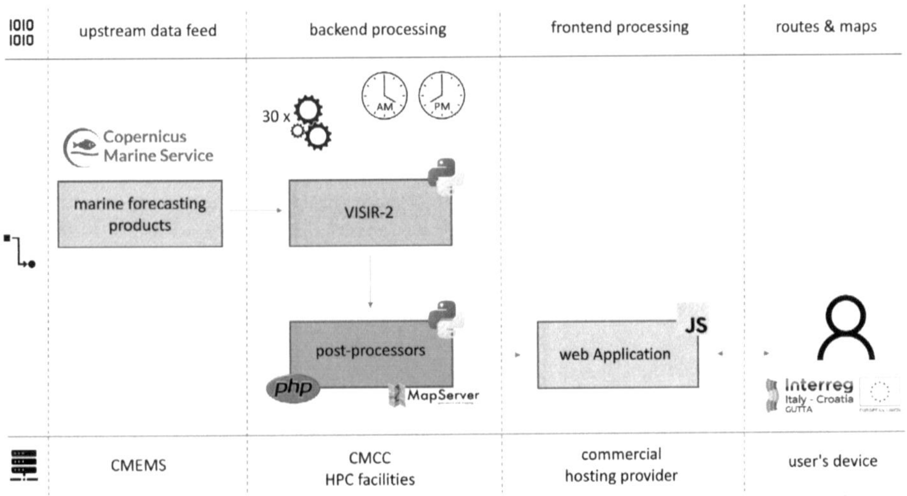

Figure 2: Overall architecture of GUTTA-VISIR. The upper row contains the name of the data streams, the intermediate one the software components , and the bottom one the hosting devices. The columns depict chronologically ordered processes. The logos next to the boxes refer to the programming languages they are coded in.

2.2 (a) Backend

The backend component is the operational chain routinely executed at the CMCC HPC facility, to produce all numerical and some of the graphical outputs of GUTTA-VISIR.
The HPC is based on a supercomputer ("zeus"[12]) with 348 SD530 Lenovo bi-processor nodes (for a total of 12,528 cores) all interconnected by means of an Infiniband EDR network. The HPC system has a computing power (theoretical peak performance) of 1,202 TFlops. An IBM Spectrum LSF software is used as a job scheduler.
The backend data flow is fed by the CMS model outputs of Sect.2.1.b. Then, twice a day, VISIR-2 optimal routes are computed. After VISIR-2 outputs its results, a post-processing programme is run, to prepare graphical elements (maps, s. Sect.2.1.d) for the subsequent processing steps at frontend level (Sect.2.2.b).
Three ferry engine load values are used: χ = 70, 85, or 100% of the engine shaft power. 39 departures every three hours are computed. This way, about five full days are covered since the time of the latest analysis of sea currents (more details in [10]). Each of these computations is performed for a total of 30 ordered departure-arrival harbour pairs in the Adriatic-Ionian region. They are listed in Fig.4.b. These computations are performed in parallel, each ordered harbour pair being processed at a distinct node of the HPC facility. In total, every calendar day, more than seven thousand route computations are completed and made publicly available.

2.2 (b) Frontend

The frontend component is the operational chain routinely executed at the CMCC HPC facility to transform the graphical elements from the backend into files used for interactive mapping. The frontend also includes a web application, to consume and visualise both this data and those numerical outputs (route waypoints and route metrics) coming directly from VISIR-2. The web app adopts a responsive layout but, given the large number of details and interacting options, its use on either a desktop computer or a tablet is recommended. More technical features about the frontend can be found in [10].

3. RESULTS

This section presents some results from the GUTTA-VISIR service. It first provides a case study (Sect.3.1) and then an assessment of the first eleven months of operation (Sect.3.2).

3.1 CASE STUDY

[12] https://www.cmcc.it/super-computing-center-scc

An example of how the GUTTA-VISIR web app looks like is shown in Fig.3. It is structured into a header menu, a sidebar, a body, and a footer. The header menu contains, on the right-hand side, links to other pages of the application (Home, About, Help, Feedback, News). The left sidebar contains the logo of the GUTTA project, along with several options which need to be selected or modified by the user to visualise, in the web app's body, the desired results. It also contains the information about the time, both the wave and current forecasts were issued. The footer contains the logos of the GUTTA project partners and the funding information.

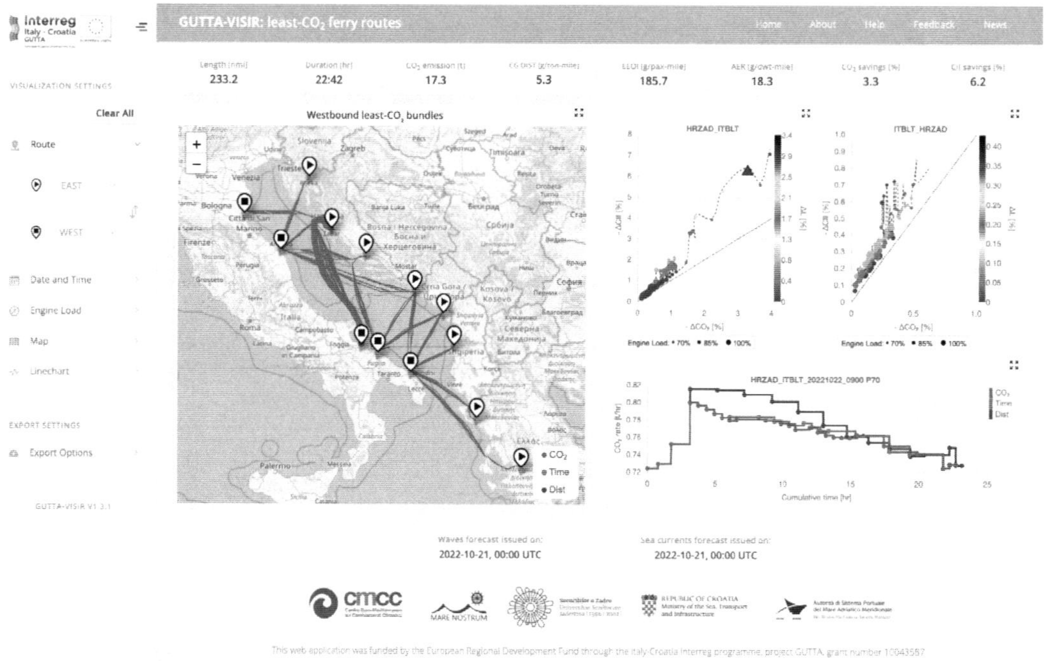

Figure 3: GUTTA-VISIR web app, upon selection of WEST in Route and click on a specific element of the Brindisi-Patras route bundle. The thickest route (Zadar-Barletta) shown in the map corresponds to the largest marker in the left scatter plot with CII vs. CO_2 savings and to the CO_2 emission profiles in the bottom right corner.

The sidebar includes tools to select both the route and its departure date and time, as well as the ferry engine load factor and some visualisation and export options. These selections affect what is displayed in the body. However, users may interact with the web app directly from the body, by selecting the end ports of the route from the map. It is possible to choose among 30 oriented routes, corresponding to either East- or Westbound sailing between 15 couples of harbours. Either individual optimal routes (like what shown in Fig.1) or bundles of routes with varying departure time and engine load (as in Fig.3) can be shown.

The routes can be explored not only in their geographical aspects, but also through some key metrics relevant to decarbonisation. They are reported in the upper part of the body (right below the web app's header menu) and include route length and duration, total CO_2 emissions, and several carbon intensity indicators (CII). The latter two metrics will be the topic of a systematic assessment done in the subsequent section (Sect.3.2.b.)

The body of the GUTTA-VISIR web app includes a section with a scatter plot of CII vs. CO_2 savings. In VISIR, both CII and CO_2 relative savings are computed with respect to the least-distance route. In the scatter plot, they are displayed for all routes between given harbours but different departure times and engine load. In [8] it is proven that the difference between these two metrics is related to the lengthening of the route due to the numerical optimization process. Thus, this plot conveys synthetic information on how effective diversions are in reducing carbon footprint of routes.

The body also includes a section ("linechart") with information along a specific route, useful to compare either the CO_2 or the CII timeline of the CO_2-optimal vs. the fastest and the shortest route between the same ports.

Finally, the CO_2-optimal routes selected for display can also be downloaded in either XML or JSON format. A massive download capacity is also available, through the CMCC data delivery system[13]. It offers an API for batch download of all routes published by GUTTA-VISIR since November 2021. They can be filtered by harbour, engine load, and date.

A structured support information for the end user is provided along with GUTTA-VISIR, in both the About and Help sections of the application. A video tutorial of the service is also available[14].

[13] https://dds.cmcc.it/#/dataset/gutta-visir/routes

[14] https://www.youtube.com/watch?v=-qORsU-Jh_8

Users feedback can be collected from a specific section of the application[15], and it was assessed in [12].
The Copernicus Marine Service has, since June 2022, listed GUTTA-VISIR as a use case[16].

3.2 SYSTEMATIC ASSESSMENT

In [8] some initial results of the VISIR-2 model in the Adriatic Sea were presented, making the point that systematic investigations on different routes and dates would be possible once an operational tool is realized. In the meantime, the GUTTA-VISIR entered operations, and has been operating for one year now. Questions like: "What CO_2 emission savings are achievable for ferries in the Adriatic Sea?", "How do they depend on the actual marine conditions or the geographic location of the routes?", "What CII savings are obtained and what fraction of the mandated ones do they represent?" can now be addressed.
To this end, all optimal routes so far produced by the operational system were analysed. This means covering the November 2021 - September 2022 time frame. The GUTTA-VISIR system has been operational since the beginning of October 2021, but the first month's data had to be pruned due to a bug in production. This led to a total of 330 days, which were analysed for the present work.
On each of these days, data for all the 30 (=15*2) oriented routes were collected, and just routes with departure times at 0:00, 3:00, 6:00, and 9:00 UTC from each port were considered. This was done for keeping in the dataset just voyages reaching destination within the same calendar day of their start. This constraint was applied in view of the subsequent analysis of Sect.3.2.b., where the CO_2 savings will be related to daily-average sea conditions. However, in GUTTA-VISIR there is no limitation regarding routes extending over multiple days.
The CO_2 savings were averaged over the considered departure times, resulting in a dataset of N_1 = 9,900 (=330*15*2) entries. Each of them was augmented with information about the average sea conditions. This was computed from daily averages of CMS analysis fields (Sect.2.1.b) for the specific date. They were spatially averaged in the subregion the route belonged to. The association between routes and subregions is provided in Fig.4.b.

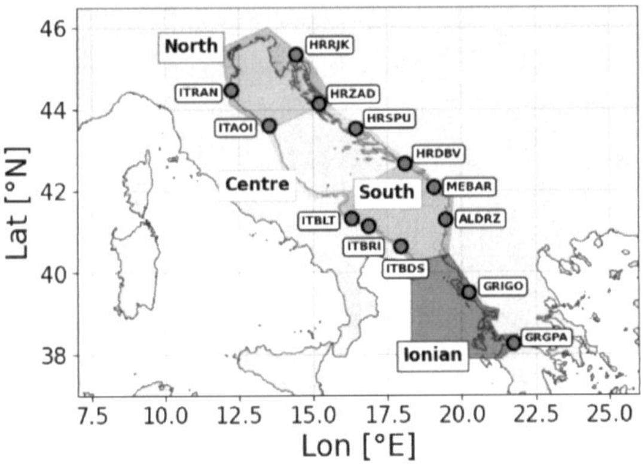

a)

North
ITRAN-HRZAD
ITAOI-HRRJK
ITAOI-HRZAD

Centre
ITAOI-HRDBV
ITAOI-HRSPU
ITBLT-HRZAD
ITBRI-HRSPU

South
ITBRI-HRDBV
ITBRI-MEBAR
ITBRI-ALDRZ
ITBDS-HRDBV
ITBDS-MEBAR
ITBDS-ALDRZ

Ionian
ITBDS-GRIGO
ITBDS-GRGPA

b)

Figure 4: a) The four subregions (North-Centre-South-Ionian) of the Adriatic-Ionian region considered in this work. The considered harbours are marked by grey circles and labelled with their international seaport codes. b) Association of the 15 GUTTA-VISIR routes (just Eastbound ones listed) to the subregions of a).

For this work, we restricted the analysis to the GUTTA-VISIR routes relative to an engine load factor χ = 70%.
The CO_2 savings were assessed in relation to both date of navigation and marine conditions (Sect.3.2.a) and sea domain (Sect.3.2.b).

3.2 (a) Role of marine conditions

The timeseries of the CO_2 savings for all routes of the region of interest are displayed in Fig.5.

[15] https://www.gutta-visir.eu/other/feedback
[16] https://marine.copernicus.eu/services/use-cases/gutta-visir-decision-support-system-ferries-adriatic-sea

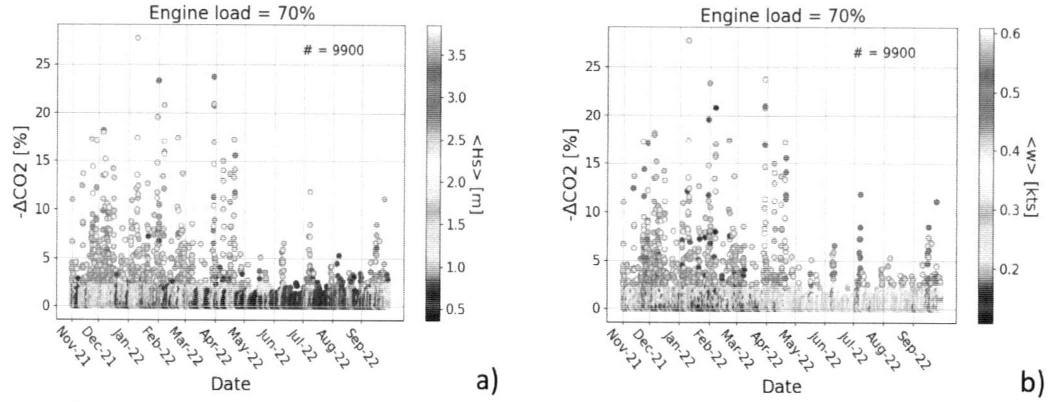

Figure 5: Time series of the CO_2 savings with marker colour representing a) significant wave height; b) current magnitude. Dataset with N_I = 9,900 entries used.

The largest savings are seen in the December-May period and can even exceed 20%. The summer values are on the average smaller, but there are two dates with savings exceeding 10%. The significant wave height always exceeded three metres whenever the savings were larger than 20%. However, during those events, the magnitude of the sea surface current could be any.

To investigate more directly the role of both waves and currents, they are used as independent variables in Fig.6's panels.

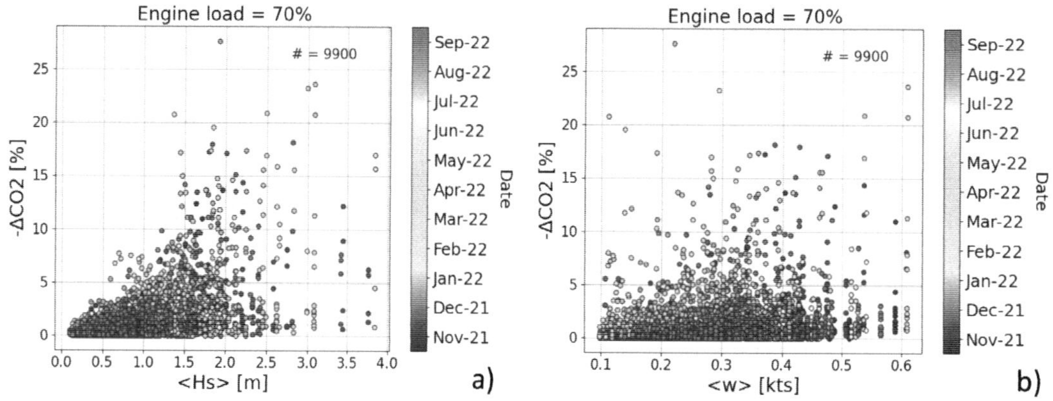

Figure 6: CO_2 savings as functions of a) significant wave height; b) current magnitude, with related date coded in the marker colour. Dataset with N_I = 9,900 entries used.

It is seen that the CO_2 savings increase their variability as the significant wave height $<Hs>$, averaged in the related subregion, increases. No similar effect is seen in relation to the current magnitude $<w>$. From Fig.6.a, the maximum CO_2 savings can be related to the sea state through the approximate relationship:

$$max\{-\Delta\ CO2\ [\%]\} \simeq 6 < Hs\ [m] > \qquad \text{(Eq.1)}$$

However, it should be stressed that the actual CO_2 savings can be any positive value smaller than what predicted by Eq.1

3.2 (b) Role of sea domain

The Adriatic-Ionian region includes both the Adriatic Sea, a marginal sea of the Mediterranean Sea with a Northwest to Southeast elongation, and the Northern Ionian Sea, a geographic sector we here defined to cover the marine domain between the Otranto Strait and the Patras Gulf.

The sea state in the Adriatic-Ionian region is characterised by a strong spatial inhomogeneity. Both the typical and extreme wind-waves are larger in the southern part of the domain. The median value of the typical waves ranges from 0.25 to 1.5 m in summer and from 0.5 to 3 m in winter, with a clear Northwest-Southeast gradient [2]. The circulation in the Adriatic is characterised by intense current segments in its western side, and three cyclonic gyres corresponding to the subregions

called "North", "Centre", and "South" in this paper. In autumn, there is a prevalent cyclonic circulation in the whole basin, with intensification of the three gyres [1].
For investigating geographic peculiarities in the GUTTA-VISIR data, the routes were first disaggregated on the departure times. Hence the corresponding dataset includes N_2 = 19,800 (=330*15*4) routes for each orientation (East- or Westbound). The dataset was then decomposed into the four subregions.
The largest CO_2 savings for Eastbound and Westbound routes were found in the "North" (30.4%) and in the "Ionian" (28.8%) subregions of Fig.4.a respectively. However, it is important to realise that the savings have a specific statistical distribution, as displayed in Fig.7.

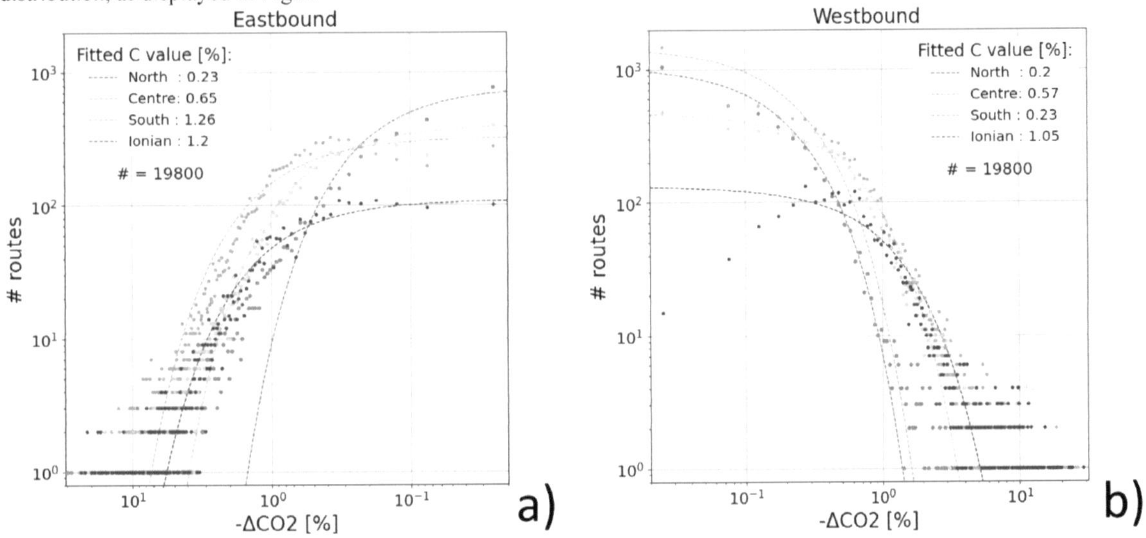

Figure 7: Statistical distributions of CO_2 savings for a) west- and b) eastbound routes. The colours refer to the subregions in Fig.4a. Markers are the values of histograms with a bin size of 0.05%. The solid lines are least-square fits of exponential functions as in Eq.2 with the *C* parameter as in legend. Dataset with N_2 = 19,800 entries for each sailing direction (East- or Westbound) used.

There, we compared the histogram data to fits of a mono-exponential function

$$p(x) = A * exp(-x/C) \qquad \text{(Eq.2)}$$

Where the quantity *C* also is the expectation value of the distribution. Routes from most subregions are exponentially distributed over three orders of magnitude. Largest deviations are seen for Eastbound (Westbound) routes of the "North" ("Ionian") subregion. In the former case, CO_2 savings larger than 0.6% are more likely than in an exponential curve. In the latter, savings smaller than 0.3% are less likely than in the exponential. This is reflected in the comparisons of the sample mean to the *C* fit parameter in Tab.2. These statistical deviations could be due to wave climate peculiarities (the "North" subregion is affected by the Bora wind; the "Ionian" routes imply sailing through the Otranto Strait, which is hardly calm) and possibly also by an inadequate spatial sampling (in our study, "North" and "Ionian" are the subregions with the least routes, cf. Fig.4b).

Table 2: Average CO_2 savings (in %), distinguished by subregions and prevalent direction of sailing. Both the fit function's (Eq.2) and the sample mean are provided.

	Eastbound		**Westbound**	
	C	sample	*C*	sample
North	0.23	0.69	0.20	0.23
Centre	0.65	1.00	0.57	0.68
South	1.26	1.28	0.23	0.69
Ionian	1.20	1.36	1.05	1.53

The values from Tab.2 are clearly lower than those predicted by Eq.1 using the average sea state values mentioned at the beginning of this subsection. Indeed, the former are mean while the latter are maximum values.
The absolute CO_2 savings of the GUTTA-VISIR optimal routes are provided in Tab.3. This information reflects both the relative savings in Tab.2 and the number of routes in each subregion, cf. Fig.4b. It is seen that a total of about 1.0 kt CO_2 would have been saved in just the Adriatic-Ionian region, during the reference period of this study of about eleven months. Considering that the total emissions along least-distance routes would have been 92.9 kt CO_2 the average savings are about 1.1%.

Table 3: Absolute CO_2 savings (in tonnes) for the eleven months of operation, distinguishing subregions, and direction of sailing. The dataset with N_1 entries per direction of sailing was used in input.

	Eastbound	**Westbound**
North	59.6	17.6
Centre	191.3	126.1
South	205.6	121.4
Ionian	139.3	163.3
subtotals	595.9	428.4
total	1,024.2	

The cost of bunker fuel, considering both the market value[17] and the allowances to be purchased in case of inclusion of shipping into the EU-ETS (Sect.1), can easily exceed 1 kEUR/tonne. If low carbon fuels are used, the cost could easily double[18]. Thus, the CO_2 relative savings in Tab.3 can have an economic value of the order of one million Euro per year. It should be recalled that they refer to just a small portion of the global ocean (the Adriatic-Ionian region), and just consider one vessel type (ferries).
Finally, we assessed the CII savings computed by GUTTA-VISIR during the first eleven months of operation. As shown in [8], CII savings are given by the sum of the CO_2 savings and the optimal route's relative lengthening with respect to the geodetic one. In view of the IMO regulation on CII mentioned in Sect.1, it would be interesting to assess if it could be fulfilled via the CII savings of optimal routes, and to what extent. To this end, we computed the distribution of CII savings exceeding a given value. They are provided in Fig.8, for the datasets of $2N_2$ routes.

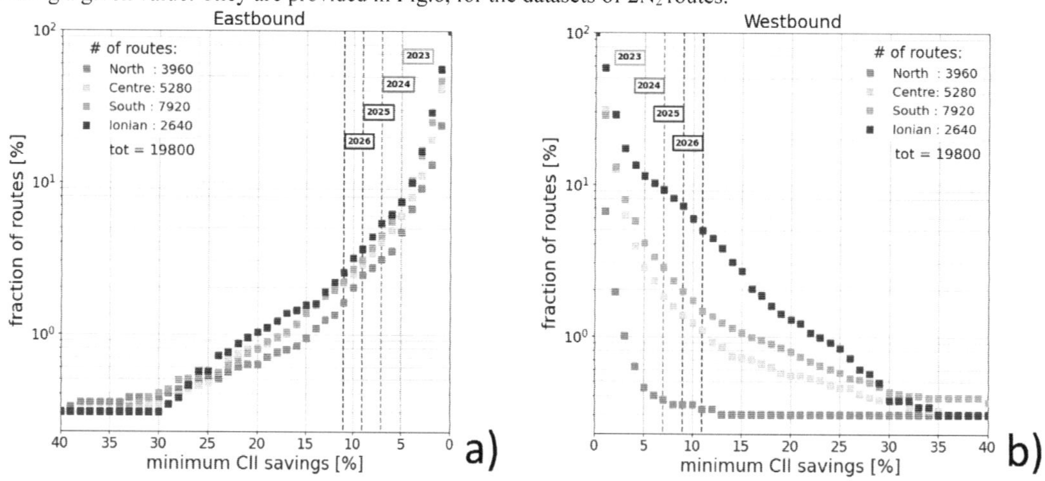

Figure 8: Statistical distributions of minimum CII savings for a) west- and b) eastbound routes. The vertical dashed lines refer to the target CII savings mandated by IMO by the years printed in the labels, taking 2019 as a baseline.

[17] https://shipandbunker.com/prices
[18] https://www.globalmaritimeforum.org/content/2022/03/Insight-brief_Methanol-as-a-scalable-zero-emission-fuel.pdf

From Fig.8 it is seen that most of the optimal routes have small CII savings, but quite large CII savings are also possible. If the Eastbound routes are considered, the four subregions have a similar distribution. For the Westbound ones instead, the fraction of "Ionian" routes with minimum CII savings up to about 30% is clearly larger than in the rest of the Adriatic-Ionian region. In Fig.8 the CII reduction targets mandated by IMO till 2026 are also displayed. For Eastbound routes, the GUTTA-VISIR savings represent between 4.7 and 7.5% of the 2023's target, and they decrease to 1.6-2.5% of the 2026's target, depending on the subregion. For Westbound routes, they represent more than 11% of 2023's target for "Ionian" routes, and this decreases to about 5% in 2026.

4. CONCLUSIONS

An operational service for computing and automatically publishing CO_2-optimal ferry routes in the Adriatic-Ionian region based on the VISIR ship routing model has been built. Such GUTTA-VISIR service has now been in continuous operation for about one year, providing open access to thousands of optimal routes, corresponding to various ports of call and departure times, daily.
This technological achievement covers a literature gap about a systematic assessment of the potential impact of ship routing on decarbonisation. By analysing 39,600 routes produced in the time between November 2021 and September 2022, we characterised the relative CO_2 savings of the optimal routes with respect to the emissions along the traditional, least-distance route. The maximum CO_2 savings on a route between given harbours are proportional to the average sea state, and we were able to derive an approximate numerical relationship between them, Eq.1. However, average savings can be much smaller and were investigated via the distribution of CO_2 savings. It was found that the largest mean savings were obtained for routes sailing in the southern part of the basin, where the typical sea states are rougher. The distribution of the savings follows quite closely an exponential one. Statistical deviations are found for specific subregions and could be due to peculiarities of the wave climate or insufficient statistical sampling. The total amount of CO_2 which would have been saved through the GUTTA-VISIR system for just ferries sailing in the Adriatic-Ionian region during the eleven-month period is about one thousand tonne CO_2.
The GUTTA-VISIR service also computed the CII savings of the CO_2-optimal routes. This metric matters in view of the IMO regulation approved at the MEPC-76 meeting. Our data show that a not negligible (up to about one tenth) fraction of the CII reduction target for 2023 could be achieved just by optimal routing. There is a sub-regional specificity in this finding, and the prevailing direction of sailing is found to matter. However, the study indicates that, even for short-sea shipping, route optimization may contribute in a sizable manner to short-term decarbonisation targets. The fraction of decarbonisation provided by routing decreases from more than 10 to about 2% of the IMO-mandated target as, in the coming years, the decarbonisation ambition rises.
To overcome some of the limitations that emerged in this research, it would be interesting to extend this work to a multi-year statistic and to a more spatially dense network of routes. The role of sea currents and engine power could be highlighted by dedicated numerical experiments. Based e.g. on [9], it is expected that ocean-going vessels would benefit even more from optimal routing. Deploying the VISIR model to other regions of the global ocean and making use of realistic representations of other vessel types would offer an objective basis for verifying and quantifying this expectation.

5. ACKNOWLEDGEMENTS

This work was supported by the European Regional Development Fund through the Italy-Croatia Interreg programme, GUTTA project, grant number 10043587. Marco Mancini is thanked for the work done to include GUTTA-VISIR into the CMCC Data Delivery System.

6. REFERENCES

1. ARTEGIANI, A., BREGANT, D., et al., The Adriatic Sea general circulation. part ii: baroclinic circulation structure, *Journal of physical Oceanography*, 1997.
2. BARBARIOL, F., DAVISON, S., et al., 'Wind waves in the Mediterranean sea: An ERA5 reanalysis wind-based climatology', *Frontiers in Marine Science*, 2021.
3. 'Network Optimization: Continuous and Discrete Models'. *Athena Scientific*, 1998.
4. 'A note on two problems in connexion with graphs', *Numerische mathematik*, 1959.
5. FRIEDLINGSTEIN, P., JOHNES M.W., et al., 'Global Carbon Budget 2021', *Earth System Science Data*, 2022, doi.org/10.5194/essd-14-1917-2022
6. MEPC.304(72) Initial IMO strategy on reduction of GHG emissions from ships. Technical Report Annex 11, International Maritime Organization, London, UK, 2018.
7. ISWG-GHG 13/3/1 Further consideration of concrete proposals on the revision of the initial strategy, and initiation of the development of a revised strategy, International Maritime Organization, London, UK, 2022.
8. MANNARINI, G., CARELLI, L. et al., 'Towards Least-CO2 Ferry Routes in the Adriatic Sea', *Journal of Marine Science and Engineering*, 2021., doi.org/10.3390/jmse9020115

9. MANNARINI, G. and CARELLI, L. VISIR-1.b: Ocean surface gravity waves and currents for energy-efficient navigation. Geosci. Model Dev. doi.org/10.5194/gmd-12-3449-2019
10. MANNARINI, G., CARELLI, L, et al., 'Operational prototype of the DSS for Eco-routes', *Deliverable 4.4.1. of the Italy-Croatia Interreg project GUTTA*, 2022. doi.org/10.5281/zenodo.6974252
11. MANNARINI, G., SALINAS M. L., et al.,. How COVID-19 Affected GHG Emissions of Ferries in Europe. *Sustainability*, 2022. https://doi.org/10.3390/su14095287
12. MIHALIĆ, M., OROVIĆ, J., et al., 'Report on GUTTA eco-routes tool user experience', *Deliverable 5.1.1. of the Italy-Croatia Interreg project GUTTA*, 2022. doi.org/10.5281/zenodo.6974258
13. OROVIĆ, J., MANNARINI, G., et al., 'Report on parametrization of vessel CO2 emissions in terms of environmental state variables', *Deliverable 4.1.1. of the Italy-Croatia Interreg project GUTTA*, 2021 doi.org/10.5281/zenodo.4778523

7. AUTHORS BIOGRAPHY

Gianandrea Mannarini is a senior scientist at CMCC. He is responsible for the development of the VISIR ship routing model and for the maritime data analysis. Gianandrea coordinated the Italy-Croatia Interreg GUTTA project, leading the development of the GUTTA-VISIR system. He holds a M.Sc. in Physics from the University of Pisa, Italy, and a PhD in Physics from the Humboldt-University of Berlin, Germany.

Lorenzo Carelli is a research associate at CMCC. He contributed to the development of the VISIR ship routing model. He holds a M.Sc in Environmental Engineering from Politecnico di Milano, Italy.

Fabio Viola is a software engineer at CMCC. He is responsible for the backend development and maintenance of the operational chains supporting several Decision Support Systems of CMCC. He holds a PhD in Computer Science and Engineering from the University of Bologna, Italy.

Megi Hoxhaj is a software engineer at CMCC. She is responsible for the development and maintenance of operational backend services. She holds a Master Degree in Telecommunication Engineering from the Università del Salento, Italy.

Matteo Scuro is a research associate at CMCC, where he mainly deals with development of Web Map Service and web-oriented applications to support the operational oceanography and the Decision Support Tools.

Saloni Kyal is a junior research associate at CMCC. She is responsible for the development of components of decision support systems and of the web applications provided for different research projects. She holds a B.Tech. in Computer Science & Engineering from Techno India University, India, and a M.Sc. in Computer Science & Engineering from Politecnico Di Milano, Italy.

Rita Lecci is deputy Director of the Ocean Predictions and Applications (OPA) Division of CMCC and responsible for the development of operational services for uptake by various user communities. She holds a PhD in Science and Management of Climate Change from Ca' Foscari University in Venice, Italy, and a M.Sc in Environmental Science and Oceanography from the University of Bologna, Italy.

Giovanni Coppini is Director of the OPA Division of CMCC. He leads the Monitoring and Forecasting Centre of the Mediterranean Sea in the frame of the Copernicus Marine Service. He is member of the Steering Committee of the CoastPredict programme of the UN Ocean Decade and coordinates its PredictOnTime core project. He holds a PhD in Environmental Science from the University of Bologna, Italy.

IMPACT OF ADVANCED HARD FOUL RELEASE MARINE COATINGS ON HYDRODYNAMICS AND SHIP EFFICIENCY

M Gaier, I Rodionov, R Ingham and **M J de Boinod**, Graphite Innovation and Technologies Inc., Canada

SUMMARY

The energy efficiency of vessels is modelled based on the roughness criteria and surface friction achieved through a novel hard foul-release (HFR) coating. The HFR coating has demonstrated the ability to repel and release biofouling in dynamic conditions, ensuring a smooth surface and low damage rates when the hull is cleaned. The HFR topcoat engineered by Graphite Innovation and Technologies (GIT) was demonstrated to combine mechanical durability with amphiphilic properties, which contributed to a modelled drag reduction in vessels compared to other marine coating types such as: silicone soft foul-release and low friction biocide-based coatings. These scientific and in-field observations support this novel hard foul-release coating as a viable tool for reducing the fuel consumption of marine vessels and meeting the international compliance standards in alignment with the new Carbon Intensity Indicator (CII), which begins in January 2023.

1. INTRODUCTION

The International Maritime Organization (IMO) has set the goal of reducing the carbon intensity of the global fleet by 40% within the next decade, and further reducing carbon intensity by 50% in total (70% intensity) by the year 2050. The Greenhouse Gas Strategy was approved by the IMO in 2018. The reduction rates are related to the baseline consumption rates in the year 2008. Short-term, mid-term and long-term measures have been established to achieve this goal. Short-term measures will begin being enforced as of the 1st of January 2023. The Energy Efficiency Existing Index (EEXI) proposed by the IMO in 2020 is a method for regulating the fuel efficiency of ships and is applicable to all vessels above 400 GT falling under MARPOL Annex VI [3].

It is a known fact, that the hull's surface roughness tends to increase due to the age of the ship, vessel operation, improper paint application, coatings defects such as sagging or improper blasting, coatings' wear-off, and biofouling, which negatively affects the ship's hydrodynamic performance. An increase in the average hull roughness subsequently leads to an increase in a vessel's fuel consumption and greenhouse gas (GHG) emissions. This phenomenon is becoming a greater issue for ships as the pressure to implement energy efficiency measures continues to grow. EEXI and Carbon Intensity Indicator (CII) measures force ship owners to reduce the operational carbon intensity of their marine assets via engine power limitation or energy saving devices [3]. Discussions on whether coatings fall under EEXI or not have been contradictory by class societies and IMO, however currently coatings are not accepted under EEXI unless sea trials are conducted before and after the application.

Reduction of external drag was shown to be realized by means of speciality coatings that would ensure inherent low friction coefficients and low fouling rates – either due to biofouling inhibition, self-release upon vessel's propulsion, or a deliberate grooming effort. The grooming of marine vessels has become a widely adopted maintenance measure that is used as biofouling prophylactics to ensure optimal vessel performance – with this, a need for both biofouling inhibiting and durable hull coatings is very high. Besides the hull, biofouling on propellers is also regarded as an important factor in with regards to a vessel's operating efficiency, impacting the propeller torque and thrust [4]. Therefore, another clear trend is the development of protective coatings to reduce biofouling and drag which accumulates on the propeller. Very little information is currently available on the impact of the propeller roughness on the energy efficiency of the vessels, especially taking into consideration the various vessel types, dimensions, geometric formatting and the interaction between the hull and the propeller performance [4].

2. EXPERIMENTS AND PROCEDURES

Mechanical performance: a list of materials science properties is used in this article to show performance, including Pencil hardness, Brush Grooming (Nylon brush), Scratch test, Wear-off test, and Pull off Adhesion.

Biofouling and foul release performance: Commercial SPC, Ice breaking, silicon soft foul release coatings, and the new developed HFR were submerged and analysed to compare their static and dynamic antifouling properties.

Surface roughness and ship resistance: A digital hull roughness gauge was used to measure the average peak roughness height at 10 randomly selected spots on each coating. The average roughness of various parts of the ship were assessed on the following segments of the model vessel's hull: flat bottom, vertical sides (port and starboard side), bow, stern areas.

Computational Flow Dynamics (CFD) simulation: A CFD methodology was developed to investigate the effect of heterogeneous distributions of hull roughness of a model vessel on its drag and energy efficiency signatures. The model showed the impact of various heterogeneous hull roughness conditions (smooth/rough) and effects on ship's resistance.

Fuel efficiency calculation: The results from the speed trials according to ISO 15016:2015 were reported by Lloyd's Register and used to determine the change in shaft power readings as affected by the newly applied HFR coating, in comparison to the readings taken from the vessels coated with other model marine anti-fouling coatings. From the relationship between shaft power requirement and fuel consumption, the resulting effect of the HFR on the fuel efficiency of the vessel was calculated. The resulting effect of the HFR on the case vessel was then compared to a theoretical hydrodynamic prediction model that was conducted using the skin friction results from a fully turbulent flow channel. The contribution of the frictional resistance to the total resistance of the vessel was calculated to determine the resulting impact of the coating's skin friction on the effective power of the vessel through the following equation:

$$\textit{\% Change in Effective Power (Fuel Efficiency)} = \frac{c_{T_{rough}} - c_{T_{smooth}}}{c_{T_{smooth}}} \times 100$$

CII calculation: By factoring in the ship type, operational profile, distance travelled, and average fuel consumption, both the reference and the attained Carbon Intensity Indicator (CII) can be calculated for any vessel over with a deadweight tonnage over 5,000GT. Once the CII reference and attained values are known, the CII rating can be determined for each year of a vessel's operation. This operational efficiency parameter is affected by the vessel owner's choice of surface preparation, hull coating system and propulsive efficiency. The equation below calculates the annual CII that is attained by a given ship:

$$\textit{Annual CII formula} = \frac{\textit{Annual Fuel Consumption} \times CO2}{\textit{Anual distance travelled} \times \textit{Capacity}[DWT]}$$

The equation below calculates the required CII for 2023 – 2026 relative to the 2019 reference line:

$\textit{Required CII formula} = CII_{ref} \times \frac{(100-Z)}{100}$, whereby: $CII_{ref} = a \times Capacity^{-c}$ and "Z" is the reduction factor ranging from 5.0% in 2023 to 11.0% in 2026.

The value of CII_{ref} is calculated using the "a" and "c" values which are determined by the ship type. Once the value of the required and attained CII values are known, the ratio of the Attained CII over the Required CII can be calculated to determine the ship's CII rating for that given year. The CII rating system spans from A to E in alphabetical order, with ratings A, B, and C meeting the energy efficiency and emission requirements set in place by the IMO. A vessel owner must submit a corrective action plan if a ship achieves a rating of "D" or "E" for three consecutive years.

3. HARD FOUL-RELEASE COATINGS AS A POTENTIAL CARBON INTENSITY INDICATOR IMPROVEMENT TOOL

High-performance marine coatings are attracting significant attention due to the increasing price of bunker fuel and emissions cap provided by IMO over the next few years. High performance coatings, engineered fuel, and engine enhancements all have potential to provide significant improvements to a vessel's efficiency. To achieve maximum operational efficiency, maritime vessels must not accumulate any biofouling or hull roughness over their drydock cycles. There is even a rising imperative to increase the operational availability of maritime vessels by extending the time between their docking cycles from 5 years to 7.5 years.

Grooming and cleaning techniques are viewed as cost-effective tools for pro-active biofouling control and are often recommended in combination with a mechanically durable coating [5]. Protective polymeric coatings with a biocide release function had long been developed to reduce the marine fouling rates but have shown limited performance over time throughout different fouling regions [6]. The fact that even the biocide-based antifouling coatings have fouling issues, has driven the marine coating industry to search for less environmentally invasive and more effective smart coating solutions. The Soft Foul Release (SFR) coatings are used as a "greener" alternative to the biocidal and self-polishing (SPC) technologies, which helped limit marine life adherence to the hull due to its smoothness, slip, chemical amphiphilicity, and gradual release of F-organo oils. Some variations of the SFR coatings were also developed to use the adhesion-inhibiting bio-active components to prevent long-term fouling growth.
The biocidal paints, SPC, and SFR coatings, were not specifically engineered to be groomed or cleaned on an active basis for a few reasons:

- *Lack of mechanical performance:* silicone-based foul release coatings are extremely soft, prompt to wear and damage due to any routine mechanical contact (such as fenders). SPC coatings on the other hand, must be soft to erode the surface layers and be effective against fouling.
- *Leaching of antifouling agents:* high concentration of silicone oil and biocides (salts of Cu and Zn), and adhesion-inhibiting synthetic bio-active compounds are actively washed out from the surficial layer during hull grooming and cleaning procedures.
- *Short lifespan of the antifouling paint*: by cleaning the surface, a layer of biofouling and coating is removed. For SPC coatings, this means that the coating would have a thinned layer of paint to be leached. For a silicone based SFR, this translates to the coating having suffered micro-scratches which becomes susceptible to faster fouling. Forces required to remove fouling, especially barnacles, from a biocide-based coating require vigorous cleaning that leads to coating loss and damage to the surface.
- *Environmental risk:* Chemical pollution due to release of antifoulants from ship hull coatings, which may be exacerbated by in-water cleaning (this is especially a concern when in port).

Mechanical Performance:

Understanding the shortfalls of antifouling coatings currently available on the market, Graphite Innovation and Technologies Inc. (GIT) has recently patented the HFR technology [7], XGIT-Fuel. This technology relies on the patented polymeric amphiphile blends upon spraying and drying and the mechanically robust finish being sleek and capable of both inhibiting the marine fouling and disrupting the biofouling adhesion strength. This technology aids in self-cleaning and ease of grooming over a vessel's lifecycle.

Scratch resistance of the HFR finish was formulated to perform over a long-term grooming-assisted life cycling on commercial cargo vessels – thus combining the benefits of the SFR's foul-release low friction and surface roughness, without impacting the health of the aquatic environment or sacrificing the coating's long-term durability. A typical lifecycle of a cargo vessel is a combination of sea faring, idling stops, and drydocking's, with any of those phases lasting from days to months, whereby the durable marine topcoat is applied to endure the intermittent marine fouling and grooming/cleaning cycles without the signs of mechanical and foul-release aptitude detriment. The mechanical durability of HFRs is viewed as a long-lasting alternative to SFRs when combined with mechanized grooming, cleaning, and roughness mitigation tools.

Figure 1 demonstrates some paint performance characteristics that were measured in comparing the HFR, SFR and SPC coating systems. By increasing the pencil hardness, scratch resistance, wear-off resistance, and coating's adhesion to substrate, it is possible to improve the grooming resistance of the coating.

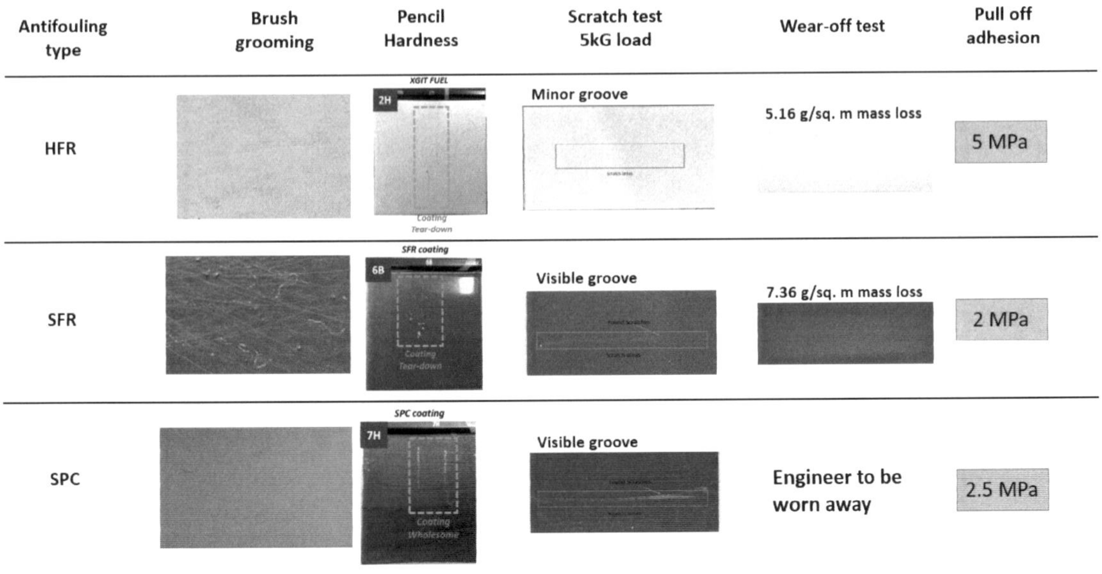

Figure 1: Anti-fouling coatings grooming and mechanical performance comparison.

Biofouling and Hull cleaning performance:

Most of the ships in operation are currently using marine coatings containing copper and other co-biocides. Due to concerns with the accumulation of copper and other biocides in many ports and environmentally protected areas, new types of biocides continue to be introduced to the market and other biocides that have been previously used are being banned. Copper and Zinc are not as effective as they once were to some biofouling species, such as Amphibalanus Amphitrite [barnacles] or Bugula neritina [tube worms], which can be viewed as the major causes of biofouling-related hull roughness and resistance increase. The main pitfall of the biocide route, however, is that these hard fouling species tend to develop resistivity to any biocide and still manage to firmly bond with the hull surface as can be observed in Figure 2.

When it comes to vessel's hull surface fouling, all the marine coatings on the market were proven to form surface biofilm, that will eventually present an anchoring profile for the subsequent layers of marine growth such as barnacles and tubeworms. Notice that with the current CII reduction cap in place, shipowners are aiming to maintain a constantly clean hull for the duration of their 5-year operational period (60 months) between dry docks. This practically means that almost all ship types (that are already well performing) will have to be groomed to have a completely clean hull, and the coating with the best CII rating for the duration of the life cycle will be the one that can better delay growth, ensure low to no damage due to grooming, and can maintain the lowest surface roughness and surface friction over time.

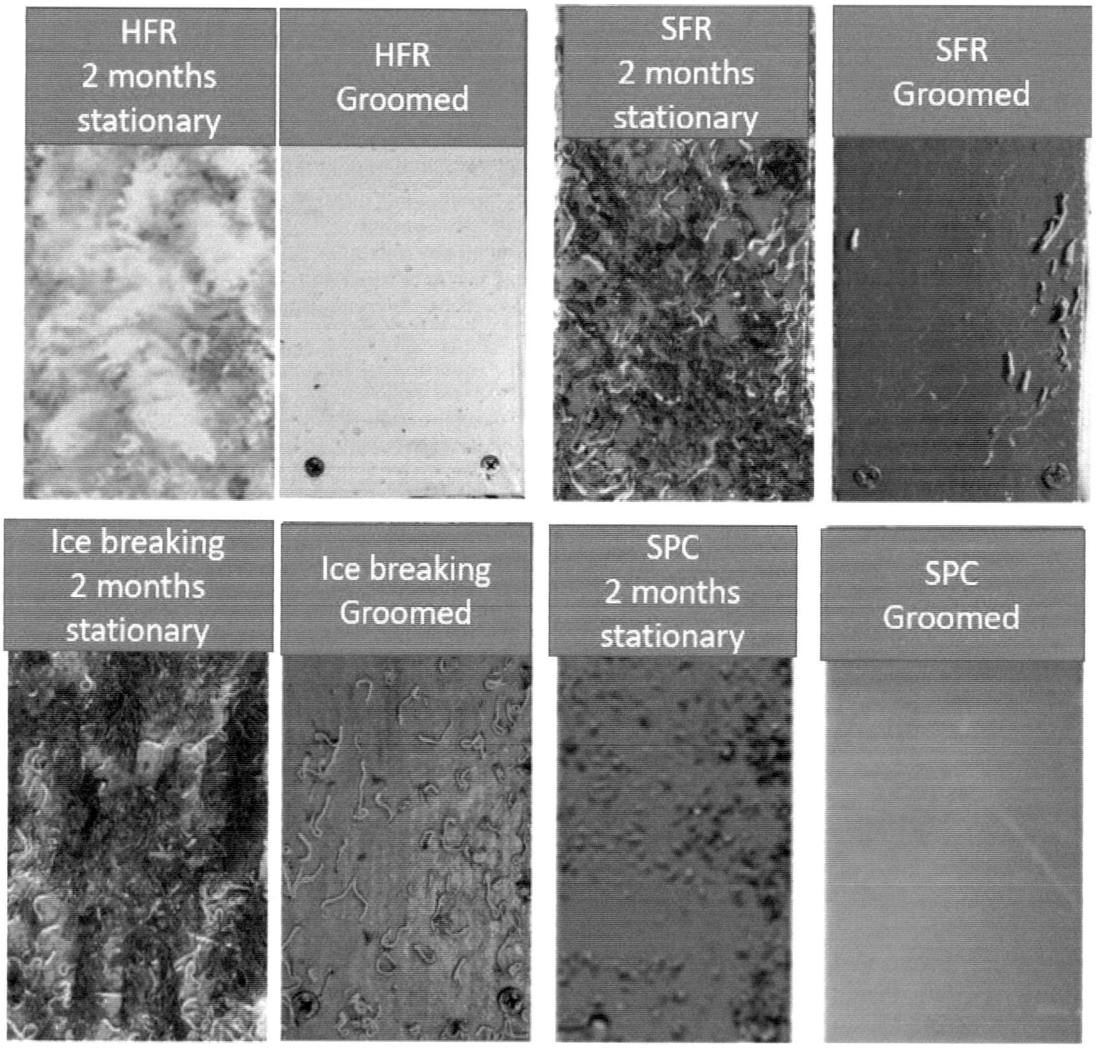

Figure 2: Antifouling marine coatings comparison. Panels have been exposed to a 2- months period stationary biofouling in tropical waters under a high fouling pressure, followed by grooming with soft nylon brushes.

Surface roughness, Surface friction and Skin friction coefficients

The prevalence of skin frictional drag of ship hulls is due to energetic, turbulent momentum transfer, which is caused by coherent structures of the boundary layer flow, e.g., hairpin vortices, biofouling accumulation, or even surface roughness. At higher ship speeds, the boundary layer becomes "thinner", therefore the amount of friction drag that is created per square metre of surface area is significantly increased and the effect of hull roughness on resistance is amplified.

Surface roughness increase due to coating roughness or biofouling accumulation will inevitably impact the ship's maximum speed and CII rating. This will be more pronounced at high speeds, where the boundary layer is the smallest. Shown below is an example of an HFR coating application in comparison to the existing SPC system in terms of their surface roughness before and after drydocking.

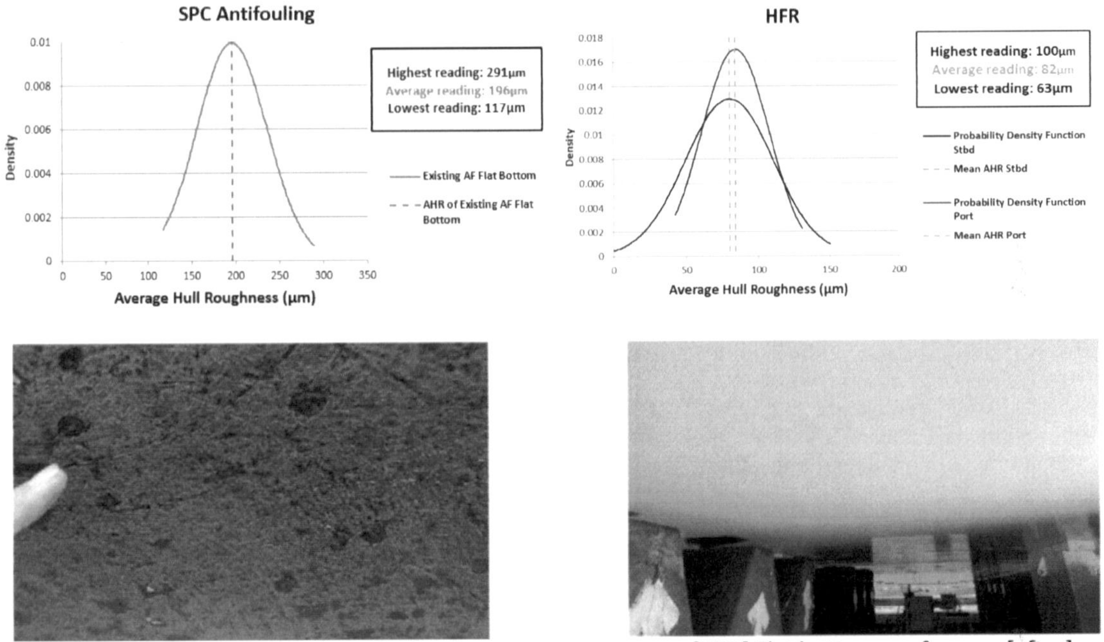

Figure 3: Average Hull roughness for SPC and HFR coatings [top images] and visual representation of the roughness [bottom images].

The boundary layer is represented by portions of transversal planes limited to the axial velocity. The roughness increase affects the boundary layer thickness around the hull in different ways, and the present results agree with previous studies. A seen below, more energy will be spent to navigate using a rough hull [SPC] than with smooth coatings [HFR].

Figure 5: Ship resistance using a CFD model with a smooth hull coating [HFR] versus [SPC] using 220 meters chemical tanker [vessel's name kept confidential as per owner request].

4. THE ROLE OF HULL COATINGS IN THE CARBON INTENSITY INDICATOR

According to the Fourth IMO GHG Study (2020), the attained CII of the global fleet has reduced by 31.8% in 2019 since 2008 which gave an average estimated improvement of 1.5% per year [8]. In MEPC 76, demand and supply based targets for CII reduction by 2030 are highlighted as 10.0% and 21.5% respectively [9]. The CII reduction targets are a strategic tool implemented by the IMO to help the global fleet reach the goal of net zero emissions. The CII rating system from A to E is directly correlated to the fuel efficiency and performance of a vessel so that both the vessel owner and its registered class society are aware of its level of performance compared to others in its category. These ratings can provide shipping companies with a measure of their fleet's environmental objectives and help them evaluate which vessels require upgrades, retrofits, etc. It is becoming ever increasingly clear that high performance low friction hull coatings are going to have an important role to play with achieving and maintaining low CII ratings.

Figure 6 below takes the reference CII value for a ~26,500 DWT gas carrier which drydocked in 2022 and applies a 2.0% CII reduction factor year over year until 2026 with 2027 assumed to follow this same reduction trend of 2.0% (a 13% total reduction from the 2019 reference value). Values of performance have been calculated using both field and lab performance data from the 4 types of coating cited in this article. As a result, the choice of the HFR coating for this vessel results in an A-rating out of dock and slowly increases to a B-rating by year 5. Both the SFR and SPC coating systems also achieve an A-rating out of dock and both progress to C-ratings by year 5. Lastly, the icebreaking coating system results in a B-rating out of dock and a D-rating by year 5. These year over year simulated CII values are a result of three main factors: achieved hull roughness and skin friction out of dock, resulting hull roughness increase from hull grooming measures, and effectiveness of the antifouling/foul release mechanism in various fouling environments. Each coating system scored differently in these categories depending on the operational profile but are generally described below:

- The CII rating of the HFR system was simulated with proactive hull grooming measures to ensure the hull maintained a smooth clean surface, knowing that the coating demonstrated the ability to be cleaned with no damage. The data for its out of dock skin friction reduction compared to the other coating systems was obtained via both field noon report data from vessels that use the technology and a fully turbulent flow channel experiment which utilized Granville's similarity law of scaling procedure to predict the coating's relative change in frictional resistance compared to a hydraulically smooth surface.

- The CII rating of the Soft Foul Release system was simulated with proactive hull grooming measures with some damage to the surface, since pure silicone-based coating systems demonstrated that they are more susceptible to damage and increased hull roughness due to proactive hull grooming measures.

- The CII rating of the Traditional SPC AF system was simulated with less of a need for hull grooming measures in the first years of operation as historical evidence suggested that most systems containing biocidal packages actively prevented the accumulation of biofouling growth within the initial months of use [10]. However, after hull grooming, this system has shown that the coating was damaged and no longer carry biocidal concentrations that are high enough to sustain its antifouling efficacy. Thus, the mechanically roughened paint surface was highly susceptible to intensive accumulation of fouling, which over time resulted in low performance.

- The CII rating of the Icebreaking system was simulated with hull cleanings as the coating type demonstrated no antifouling properties under dynamic conditions and did not have any other methods of managing accumulated biofouling growth. The icebreaking coating also showed that after idle periods or dynamic conditions, the biofouling growth could not be removed with traditional hull grooming methods, leading to a large increase in hull roughness and thus the highest CII rating over time.

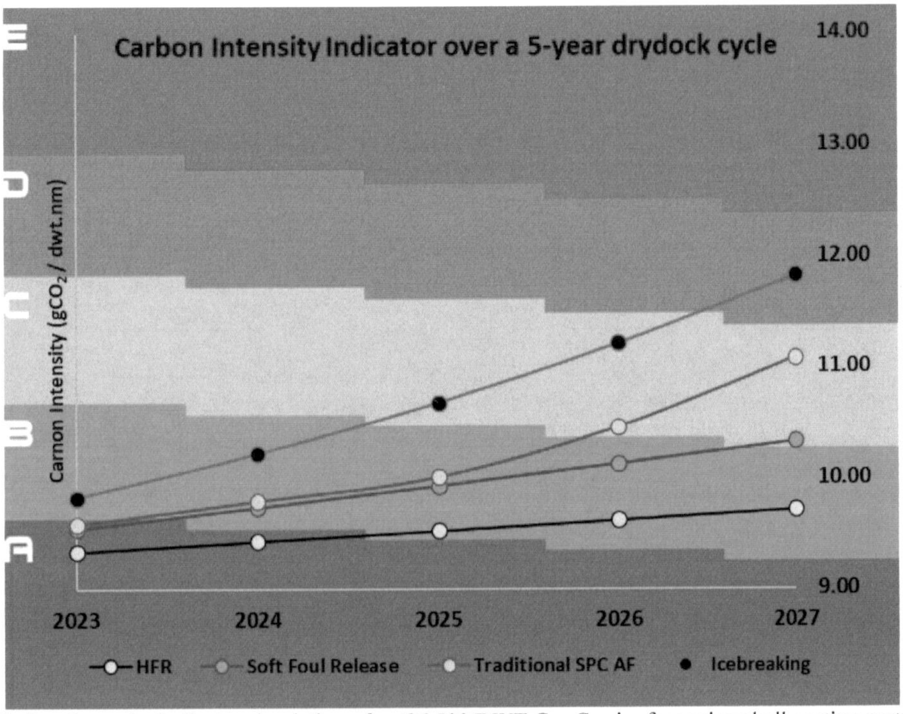

Figure 6: Carbon Intensity Indicator simulation of a <26,500 DWT Gas Carrier for various hull coating systems over a 5-year drydock cycle (2023 – 2027).

Table 1 assumes the same annual distance travelled (km) and operational profile as the base year (2022) for this gas carrier and is an example of another method to show the change in CII rating over time for each coating system. Vessel owners can utilize this table format in combination with their vessel specific characteristics to make educated decisions on the hull coating choice for the surface preparation and hull maintenance practices over their ship's drydock cycle.

Table 1: Summary table of the Carbon Intensity Indicator simulation of a <26,500 DWT Gas Carrier for various hull coating systems over a 5-year drydock cycle (2023 – 2027).

Coating System Type	Initial	2023	2024	2025	2026	2027 (est.)
Full blast, HFR with proactive hull grooming measures	10.37 (B)	9.34 (A)	9.43 (A)	9.52 (B)	9.62 (B)	9.72 (B)
Full blast, SFR with proactive hull grooming measures	10.37 (B)	9.54 (A)	9.73 (B)	9.93 (B)	10.13 (B)	10.33 (C)
Full blast, Traditional SPC AF with reactive hull cleaning measures	10.37 (B)	9.58 (A)	9.79 (B)	10.01 (B)	10.46 (C)	11.08 (C)
Spot blast, Icebreaking with reactive hull cleaning measures	10.37 (B)	9.82 (B)	10.22 (B)	10.67 (C)	11.21 (C)	11.83 (D)

5. HULL INSPECTIONS AND GROOMING

Trade, vessel usage, and faring routes are all critical components when it comes to maintaining a hull free of biofouling and retaining an optimal CII rating. Figure 7 demonstrates an image of the hull inspection of a model vessel [name kept confidential per owner's request], coated with the HFR product while having undergone regular operation for 6 months. An important note would be that the vessel's in-water faring activity was 87% at the time of the inspection at an average cruising speed of 9 knots under medium fouling pressure during the summer months. The hull showed no signs of fouling and maintained it's out of dock operational efficiency.

Figure 7: Hull inspection of HFR after 6 months in-service. The vessel was operating at its standard operational profile.

Figure 8 shows the HFR versus Ice Breaking paint after being alongside for 3.5 months in tropical waters in a high fouling region. The vessel currently has HFR, and Ice Breaking paint applied to its hull and showed a large amount of fouling on the Ice breaking paint section (100% coverage), while only having the juveniles' barnacles on HFR section. Both coating systems are shown before and after they were groomed. The HFR test area was cleaned using nylon brushes while the icebreaking paint area required three separate hull cleaning sessions using metal wire brushes to achieve somewhat of a cleaner hull surface than the original. Notice that with ice breaking paint, the barnacles are not fully removed, and the remaining "footprints" of the organisms are still present after the hull cleaning.

Figure 8: Hull inspections before and after hull grooming. HFR and Ice breaking paint comparison.

6. CONCLUSIONS

A slippery, smooth, mechanically durable coating class that is hard and easily cleaned was developed to reduce marine fouling and increase vessel performance and efficiency. The novel hard foul release coatings were tested side by side with the conventional soft foul release coatings, abrasion resistant hard coatings and traditional SPC antifouling coatings. Modern coatings with superior mechanical properties and the ability to maintain low roughness, zero to minimal damage, and low skin friction drag can maintain an optimal CII rating if the correct hull maintenance plan is implemented according to the vessel routes, trade, and activity. Hull coating systems can be the difference between an A and a D CII rating for well performing vessels and for poor performing vessels it can mean the difference between a B/C and an E rating, warranting the need for a corrective action plan from the vessel owner. The CII calculation results show that the last two years of the antifouling system's lifetime are critical for the vessel to maintain a good CII rating and operational efficiency, while the current SPC systems show little to no performance benefit in the last two years of the vessel's operation.

7. ACKNOWLEDGEMENTS

Acknowledge to all the colleagues at Graphite Innovation and Technologies Inc. Also, to Dr. Geoffrey Swain and Dr. Melissa Tribou from Corrosion and Biofouling Control (CCBC) at Florida University of Technology for some of the results on biofouling performance.

8. REFERENCES

1. Schultz, M., 'Effects of coating roughness and biofouling on ship resistance and powering', Biofouling, 2007.
2. Ytreberg, E., 'Valuating environmental impacts from ship emissions – The marine perspective', Journal of Environmental Management, 2021.
3. IMO, 'Further shipping GHG emission reduction measures adopted'. 2021
4. Sezen, S., Uzun, D., Ozyurt, R., Turan, O., Atlar, M., 'Effect of biofouling roughness on a marine propeller's performance including cavitation and underwater radiated noise (URN)', Applied Ocean Research, 107, 102491, February 2021.
5. Tribou, M., Swain, G. The use of proactive in-water grooming to improve the performance of ship hull antifouling coatings. *Biofouling*, 26:1, 2010.
6. Catão1, E.C. P., et al. 'Shear Stress as a Major Driver of Marine Biofilm Communities in the NW Mediterranean Sea, Front. Microbiol., Sec. Aquatic Microbiology, 31 July 2019.
7. M. Gaier, M. AlGermozi, and I. Rodionov, "Composition for a coating. coatings and methods thereof," , Internation Patent WO2021/226699 A1, Graphite Innovation and Technologies Inc., May 2021.
8. International Maritime Organization (IMO), 'Fourth IMO Greenhouse Gas Study 2020', August, 2020.
9. THE MARINE ENVIRONMENT PROTECTION COMMITTEE', 'Resolution MEPC.338(76) 2021 GUIDELINES ON THE OPERATIONAL CARBON INTENSITY REDUCTION FACTORS RELATIVE TO REFERENCE LINES (CII REDUCTION FACTOR GUIDELINES, G3', 17 June, 2021.
10. Tribou, M., Swain, G. The effects of grooming on a copper ablative coating: a six year study. *Biofouling*, 33:6, 2017.

9. AUTHORS BIOGRAPHY

Marciel Gaier, PhD is a Chief Technical Officer and Technology visionary at Graphite Innovation and Technologies.

Ilia Rodionov, PhD holds the current position of Product Formulation Supervisor at Graphite Innovation and Technologies. His mission is the development and scaling up the sustainable polymer coatings formulas and securing the IP. His previous experience includes a PhD in polymer material chemistry and biotech devices engineering, his applied and academic research was reflected in the authorship of 10 peer-reviewed publications and 3 inventions.

Ryan Ingham holds the current position of Energy and Efficiency Lead at Graphite Innovation & Technologies. He is responsible for hull performance analysis and methodologies, life cycle assessment and commercial strategy for EEXI & CII. His previous experience includes a master's thesis on the topic of life cycle assessment of marine coatings.

Michael Jacot de Boinod, BSc holds the current position of Data Scientist at Graphite Innovation & Technologies. He is responsible for managing partner datasets, creating custom analysis tools, and visualizing performance results. His previous experience includes a bachelor's degree in Physics and Computer Science, as well as experience developing testing software for energy infrastructure.

TAKING THE CREW ALONG ON THE VOYAGE – SHARING OPERATIONAL INSIGHTS WITH THE CREW REDUCES CARBON FOOTPRINT AND INCREASES FUEL EFFICIENCY

M Köpke, Hapag-Lloyd AG, Germany

ABSTRACT

The reduction of fuel oil consumption and CO2 emissions is an economic as well as environmental target for each shipping company. One, if not the most, crucial element in the process of efficient voyage execution is the crew. To achieve a higher level of transparency and mutual understanding of operational performance Hapag-Lloyd introduced a variety of awareness initiatives. In this continuous process of fostering awareness automated feedback to vessel crews for all 250 operated ships were developed. This automated crew feedback displays the results of the vessels' actions and reports. Each reported value is compared to an expected value, which is based on a virtual model of each individual ship. The crew is immediately aware of reporting errors, measurement errors and possible excess consumptions. This paper elaborates on the background of the feedback, and crew awareness and will analysis the impact on data quality, vessel operation, and efficiency. Further, the bottlenecks and obstacles will be analysed which are hindering the crew to gain from such a tool. The feedback to the crew module was a development with the innovation project ShippingLab.

Keywords: containership, energy efficiency, low emissions, operations, crew awareness, digital twin, voyage execution, ShippingLab

1. INTRODUCTION

Decarbonisation is one of the biggest challenges for the shipping industry today and the coming years. The shipping industry has clear GHG reduction targets released by IMO 1 and EU 2. Beyond these mandatory reduction schemes, shipping lines, as Hapag-Lloyd, published even more ambitious targets towards decarbonisation of shipping 3.For new build vessels, regulations exist, e.g., with the Energy Efficiency Design Index 4. Certainly new, carbon free, propulsion technologies are required. However, there is still a large existing fleet, which is to a certain extent modernised towards operation that is more efficient. Nevertheless, these vessels operate on fossil fuels. Therefore, the vessel operation itself needs to be as efficient as possible. IMO addressed this with various levers, such as the introduction of the Ship Energy Efficiency Management Plan (SEEMP), the Data Collection System (DSC) and the Carbon Intensity Index (CII) 5,6. The latter will have significant impact on the current fleet, by increasing the required energy efficiency on a yearly level. To reach these targets speed instructions will have a significant impact, however, also the efficient voyage execution is even more in focus than before.

Like IMO's initiatives the EU introduced the Monitoring, Reporting and Verification (MRV) scheme, where shipping companies report their emissions in European waters. EU already announce that an Emission Trading System will follow, where shipping companies need to pay for their emitted GHGs.

Together with IMO's DCS this gives a new importance to the quality of reported operational data from ships.

This paper will discuss the role of crew awareness and how behaviour changes can be successfully achieved. A combination of trainings, software and transparency initiative will be introduced, among other the roll out of a fleet wide performance feedback to crew platform. To begin the relevant Key Performance Indicators (KPI) will be introduced. Followed by an analysis of the impact on performance and a discussion of the obstacles. The introduced crew feedback module is a development within the ShippingLab. ShippingLab is a non-profit innovation and project partnership in Blue Denmark 7.

2. KEY PERFORMANCE INDICATORS

Operational data quality and the fuel consumptions are in the focus of shipping companies as well as authorities and policy makers. Therefore, it needs to be determined what is needed by the crew to improve theses.

The fleet operated by Hapag-Lloyd consists of more than 250 container vessels varying in size from 350 to 19,000 TEU. Further, approximately one-half of the fleet is owned, the other is chartered. This means communication, on board systems, technical equipment and training of the different vessels and crews varies. For operational data reporting that means the traditional noon report is the one common ground all vessels can provide.

To improve voyage execution and data quality the crews were made aware of their importance on board. This awareness campaign targets especially Captains, Chief Officers, Chief Engineer and 2nd Engineer, since they hold the main responsibilities for the nautical and technical speed and power management of a ship. Furthermore, they are also responsible for the operational reporting. To achieve the targets towards higher data quality and more efficient voyage execution the following awareness initiatives were introduced:

- On shore monitoring and voyage execution support,
- Regular newsletters,
- Energy efficiency workshops,
- Remote trainings,
- E-learning,
- De-briefing meetings,
- On-board visits, and
- Performance feedback to the crew.

These initiatives were subsequently introduced. At all times one focus was however, to keep the crews informed about developments in monitoring and energy efficiency. Tools, which support decision making on board, e.g., trim optimisation tools, were never just sent on board, but always accompanied by more comprehensive initiatives, which include clear communication what the initiative is about and what is expected, brief but meaningful instructions, trainings, and transparency of the resulting impact. For some initiatives not only the vessels and crews were in focus, but also the shore organisation. Such as for the example of trim optimisation the stowage planning offices were trained to enable the vessels to optimise their trim.

To close the loop of voyage optimisation, operational reporting and performance analysis an online feedback to the crew was developed and rolled out. In case of Hapag-Lloyd all vessels report in the same tool to have comparable and unified data for each vessel. Hence, in the first roll out of the feedback to crew all analysis rely on manual noon data. Following requirements to the feedback to the crew were established:

- Visualise reported figures to the crew,
- Enable the commands to see their impact of their own actions,
- Quantify data quality and vessel performance, and
- Compare themselves to (anonymised) peer vessels / sister vessels and of course.

The operational data quality is shown as KPI as well as percentage of errors, which triggered a plausibility check. Further, the respective errors for each day are displayed.

The main engine's total consumption as well as the specific fuel oil consumption are displayed. The consumptions are shown in relation to the expected consumption based on fuel tables and shop test, respectively. Furthermore, speed-consumption data in relation to weather and a load diagram are available in order to support the vessels crew in their voyage planning.

The actual and the expected boiler consumptions are displayed.

In order to stimulate the professional pride and completion the KPIs of the anonymised sister vessels are available to the crew.

In general, most of these data was available to the vessel command also before the crew feedback platform was rolled out; however, on the platform also the expected values are shown. This help to identify reporting, but also senor related issues.

By providing sister vessels' KPIs the board command can also see what is possible regarding efficient operations on other vessels.

3. IMPACT REVIEW

Since the afore mentioned awareness measures were introduced over a longer time period and some are enabler rather than generating actual savings a quantification of the total impact of awareness is at this stage not possible. However, the impact of single initiatives can be quantified. For this paper trim optimisation and the feedback to the crew are investigated further.

The impact review is done on two levels, on one side the perception (qualitatively) of the implementation of the crew feedback platform is evaluated. On the other side, the implementation is quantitatively evaluated regarding the KPIs.

4. QUALITATIVE REVIEW

For the review of how the crew feedback was perceived on board a survey form was send to vessels. The answering of the survey was anonymous in order not to bias results. About 20% of the fleet filled out the survey form. About 7% of the vessels wrote back that they had technical issues to fill out the form and the rest did not respond.

The vessels were asked questions in distinct categories such as frequency of usage, utility, system performance, practicality and completeness of the platform.

The answers of the survey show that the general perception of the feedback on board is incredibly positive. The vast majority of 96% answered that they deem the feedback to crew as useful and 2/3 of the vessels answered that they use the tool at least once per week, see Figure 1 and Figure 2.

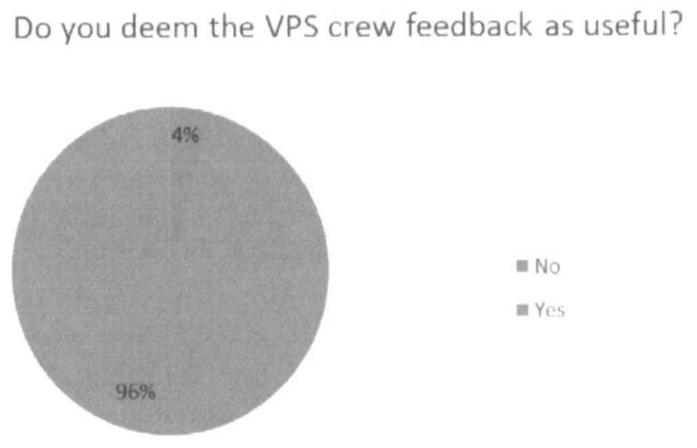

Figure 1: Usefulness

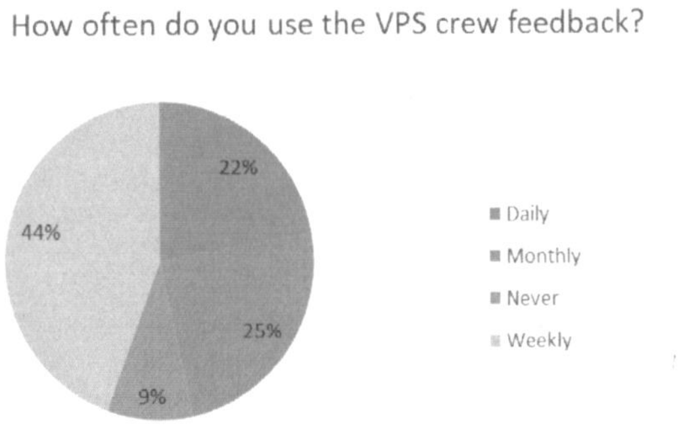

Figure 2: Usage frequency

Furthermore, most of the vessels stated that the crew feedback improves the overall voyage execution as well as quality of the reported data, see Figure 3 and Figure 4.

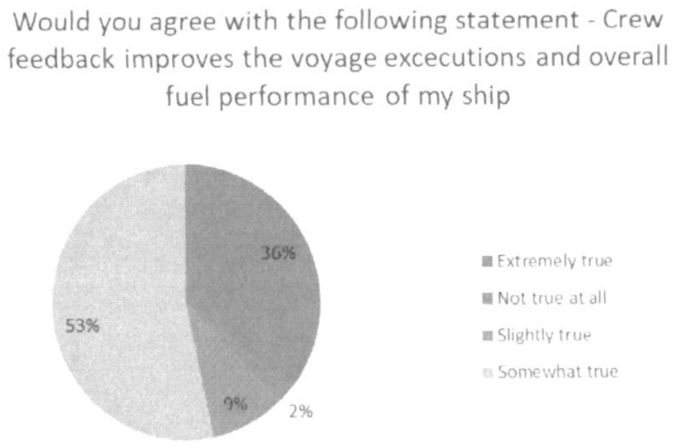

Figure 3: Perception of voyage execution and overall fuel oil consumption improvement

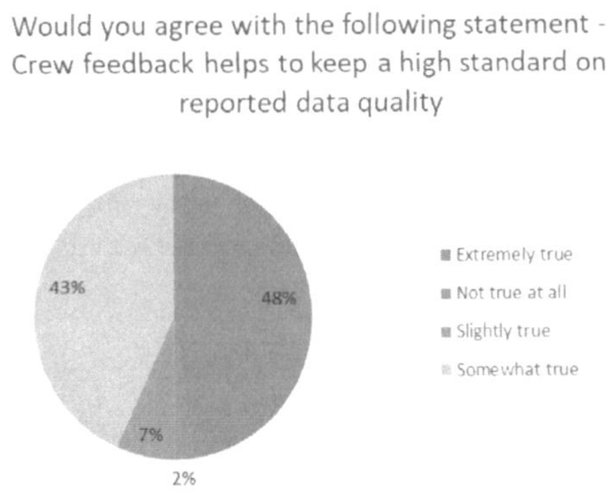

Figure 4: Perception of operational data quality improvement

One assumption behind the introduction of the feedback to crew model was that the board management would appreciate the transparency and perceive it as additional motivation. The vessels were asked for brief testimonials how they perceive the feedback. In general, these statements reflect that the crew appreciates the feedback to crew module. Exemplarily, some statements are shown below; please note that two Captains sent their testimonial via email and therefore, with their consent, are not anonymised:

"Please be informed that this performance monitoring tool is very useful in order for us to improve our machineries performance in general and to rectifying the warnings and errors as well." – Anonymous

"Mainly for curiosity, to compare our performance with other ships." – Anonymous

"The best is speed and consumption for future voyage calculation." – Anonymous

"Monitoring for highest amount of errors and analyse where are they coming from. Analysing performance of Main Engine and comparing with benchmark." – Anonymous

"We can see performance of vessel during voyage, and we can see various graphs and it help us to identify performance and maintenance issues". – Anonymous

"The automated crew feedback software gives us a feedback of the quality of our reported data and proves to us that our efforts in collecting and reporting these data produce actual results. Comparing our data to peer values and to anonymized sister vessels can help us find abnormalities in the data reported, enabling us to improve the quality of our noon reports or even vessel performance as a whole. With a generally increasing workload of documentation and reporting on board, it is nice to have a tool that uses data, which we already provide anyway, without creating additional work." – Dennis Schwartz, Master Mariner

"For several years we have been using the same reporting tool on all ships in the HL fleet to report a large number of parameters from daily ship operations, e.g., positions and times at the start/end of the voyage, midday position, miles travelled, observed weather data, consumption of main engine, auxiliary diesel and boiler etc. The "Fleet Analytics" team on land evaluates this data. The positions are forwarded to our weather provider for weather routing. With the crew feedback, we get a visualisation on-board. This enables to compare data we previously entered voyages and as well as corresponding values of the sister ships. The latter is particularly important in my service trade area with five sisters in the same service. The tool also enables us to check the quality of the data entered. In the medium and long term, even possible technical defects can be identified that occur gradually and are therefore not noticeable in daily on-board operation (e.g., defective valves that cause higher boiler consumption)." – Björn Kropp, Master Mariner

5. QUANTITATIVE REVIEW

The quantitative analysis investigates if there are measurable improved since the introduction of the crew feedback. For this, a subset of 38 owned and by Hapag-Lloyd managed vessels is analysed. This subgroup of vessels is used to ensure the vessels were continuously in the fleet for the whole period that this paper investigates, which could not be guaranteed for chartered vessels.

The focus of the quantitative analysis in this paper is on the data quality and not the consumptions. The reason are different external parameters that have an impact on the consumptions that are beyond the crew feedback. To illustrate this the monthly average consumption of the oil-fired boiler is analysed, six months before and six months after the crew feedback was introduced. The chosen subgroup of vessels show that the boiler consumptions increased by around 10%. However, the six months after the introduction of the crew feedback are the months from October 2021 to March 2022. All analysed vessels operated in most of that period in the northern hemisphere. Therefore, it is concluded that the boiler consumptions correlate to the lower average temperatures, see Figure 5. However, even on a year-to-year comparison the consumptions slightly increase. Yet, no clear conclusion on the additional consumption could been derived. There is consensus that this increase is not cause by providing additional data to the crew.

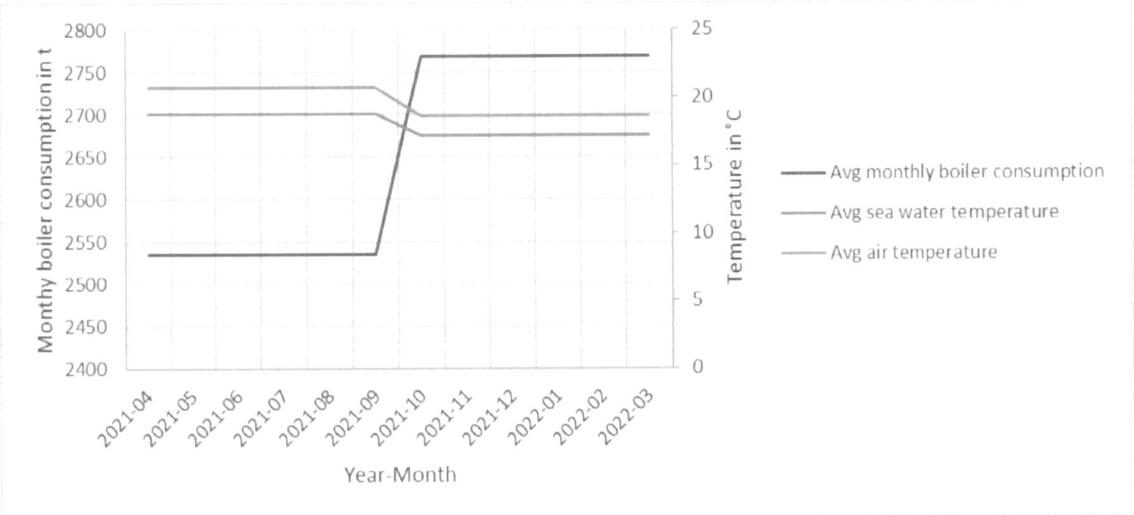

Figure 5: Monthly avg. boiler consumption at sea and sea water / air temperature

To quantify the impact on the reported noon data quality, the weekly average number of error is counted. An error in this context means that a plausibility check is triggered, and the respective reported value is out of expectations. Again, the earlier defined subset of vessels is used. For all analysed vessels the data quality improved by about 7%, i.e., 7% fewer errors are detected. However, it became also apparent that some vessels showed a decay in the data quality. It seems unlikely that by focusing on data quality the actual data quality would decrease. In this context, it should be noted that the actual usage of the crew feedback tool could not be tracked. Therefore, it is considered that the tool where data quality decreased was not in use.

When filtering out the vessels that may not use the tool the improvement on data quality is much higher. The increase of data quality is about 21%, see Figure 6.

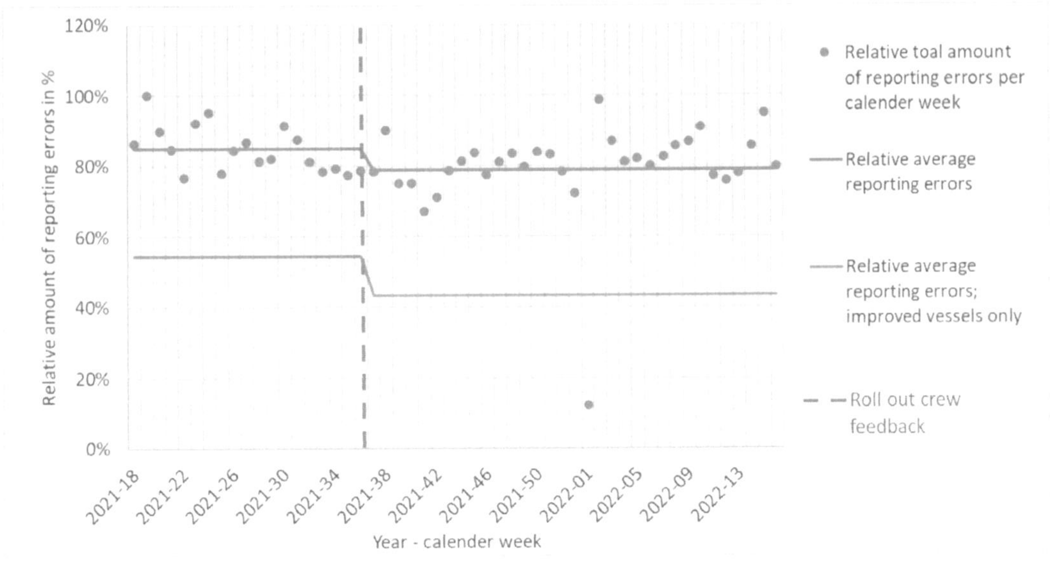

Figure 6: Operational data quality

Today, 173 out of 253 Hapag-Lloyd's operated vessels are equipped with trim optimisation tools. Among others the crew on board see the trim recommendations in a standalone tool on board for voyage planning purposes as well as in the feedback to crew tool.

Fuel savings and hence emission reduction due to trim is measured by analysing the theoretic consumptions based on the Computational Fluid Dynamics (CFD) simulation at the actual trim and a benchmark trim. The benchmark is each vessel's individual weighted average trim before the tool was installed on board and even keel for new vessels, respectively. On average trim optimisation results to a decrease of fuel consumption of about 1.5%.

7. LESSONS LEARNT

Mutual understanding of the needs on board as well as onshore should be established. This can be done with various measures. Most effect are face to face and remote trainings. Personal exchange was found as an effective measure to foster understanding and trust between stakeholders. Every initiative should be accompanied by circulars, newsletters and instructions. Ideally, there should be a balance between informing the crews and not overloading with information.

The core to the successful implementation of trim optimisation is a joint achievement by the Masters and officers on board, the stowage planners ashore, as well as training on the tools, sharing of experiences and transparency of the results. On board the crew checks the actual loading conditions, bending and shear forces and the trim optimisation recommendation and derives with a decision which trim is possible. Prior to this our stowage planners will plan the cargo in a way that the vessel is able to achieve the most advantageous trim as possible, without violating bending and shear force limits.

Achieving transparency is key to improve. Within the rollout of the feedback to crew module, some observations were made. First, about 5% of the chartered vessels have no internet access, but for emails. This could not be solved during the project even though discussing the issues with the managers/owners and hence the tool could not be rolled out to these ships.

Furthermore, about 25% of the chartered in vessels could initially not reach the webpage of the crew feedback, because it was blocked. The IT departments of the respective ship managements needed to whitelist the webpage. This was necessary for internet security reasons and company policies.

In addition, on the owned and managed vessels the crews reported at the beginning that the performance of the system was extremely low, resulting in inconvenient waiting times. In some cases, the waiting time were so frustrating that Captains would not use the tool. After some investigations with Hapag-Lloyd's IT it was found that the reason for the low performance was related to internet security settings. The settings of the involved systems could be configured resulting in much faster performance of the system.

Initially it was planned to track the usage of the crew feedback module using Google Analytics. After continuously monitoring lower than expected number of users the presumed non-users were asked for the reasons why not using the tool. It turned out that many of the presumed non-users stated that the tool is in use. This led to an investigation if the tracking of clicks in Google Analytics works sufficient. It turned out that depending on the settings of individual computers (basically cookie settings), satellite communication, company internet security settings, etc. Google Analytics would not count all click. Hence, a new tracking needs to be developed.

To stimulate the usage of the tool as well as receiving feedback by the vessels weekly newsletters were send to the fleet introducing functions and features of the tool. In addition, improvements for data quality, which could be related to the use of the tool, were communicated.

8. CONCLUSION

For successful behaviour chance one should effectively communicate what measures are introduced and why. Demands and operational requirements of vessel crews as well as shore organisation should be considered. Transparency is one of the key elements. Seeing the impact of their actions, stakeholders are more motivated to chance and to increase their efforts towards even more efficient voyage execution.

The presented analysis shows that the crew feedback tool is perceived very positively within the fleet. The increased level of transparency provides information for more efficient voyage execution, supports voyage planning and helps to identify efficient operational practises.

Furthermore, the analysis showed measureable improvement of operational data quality. By increasing operational data quality, the reliability of efficiency evaluation increases. This applies to all consumers as well as hull efficiency evaluation. For the latter, higher reliability and more accurate data lead to quicker conclusions and hence actions if a performance decrease is detected, e.g., due to biofouling.

Even though, this paper does not quantify benefits in consumptions there is consensus that voyage execution and hence consumptions are improved. This is also stated by the on-board managements.

As a next step, in order to also reach vessels with no internet access, but emails, vessels with slow internet connections and on-board managements who simply do not click the tool, regular performance reports will be send on board.

Generally, to achieve best possible exchange of information communication of all involved parties should be open and respectful communication. This supports identifying obstacles as well as best practise examples.

9. ACKNOWLEDGEMENT

The author thanks all involved Captains and Officers of the Hapag-Lloyd's operated fleet for their valuable comments and feedback. In addition, the author thanks all members of the ShippingLab innovation project for the constructive collaboration, bright ideas and developments.

10. REFERENCE

1. IMO RESOLUTION MEPC.304(72) "INITIAL IMO STRATEGY ON REDUCTION OF GHG EMISSIONS FROM SHIPS", adopted on 13 April 2018
2. EU COM(2021) 562 final, "Use of renewable and low-carbon fuels in maritime transport and amending", Directive 2009/16/EC, July 2021
3. Hapag-Lloyd Capital Markets Day 2021 presentation, 17. Nov. 2021; https://www.hapag-lloyd.com/de/company/ir/calendar-events/capital-markets-day.html#tabnav
4. IMO RESOLUTION MEPC.203(62), MEPC 62/24/Add.1, July 2011
5. ISWG-GHG 1/2, "Existing IMO activity related to reducing GHG emissions in the shipping sector", Feb 2017
6. IMO "Annual operational carbon intensity indicator (CII) and CII rating", https://www.imo.org/en/MediaCentre/PressBriefings/pages/42-MEPC-short-term-measure.aspx, April 2022
7. ShippingLab, https://shippinglab.dk/, October 2022

11. AUTHORS BIOGRAPHY

Martin Köpke holds the current position of senior manager at Hapag Lloyd's decarbonisation team, he is also secondee to the Mærsk Mc-Kinney Møller Center for Zero Carbon Shipping. Martin joined Hapag-Lloyd energy efficiency expert in December 2013. His work includes the development and evaluation of energy efficiency and decarbonisation measures.

Before Hapag-Lloyd Martin worked at Germanischer Lloyd's Research & Development unit for about 7 years. His previous experiences include operational energy efficiency, ship design optimisation, development of consultancy services and advisory of the German flag in questions of energy efficiency in international bodies.

Martin graduated as a mechanical engineer with emphasis on naval architecture.

ENERGY EFFICIENCY IMPROVEMENT FOR LARGE PASSENGER SHIP THROUGH HYDRODYNAMIC TESTING

M Pedišić Buča, Croatian Shipbuilding Corporation, Croatia

SUMMARY

A design of fully electric 105 m-long passenger ferry was developed for sea traffic between the mainland and islands on the east coast of the Adriatic Sea. It is to be the first electrically powered larger passenger ship design envisaged for this area.
Besides obvious benefits related to the elimination of emissions, the designer decided it to conduct hydrodynamic optimization in the towing tank. With such an approach, undertaking simple changes and modifications to the design lead to significant savings. Extensive tank tests were conducted at Brodarski Institute - Croatian Shipbuilding Corporation, Zagreb, Croatia, and savings are demonstrated. This paper describes the scope of the work done and provides guidelines which may serve as an orientation for the possible impact of specific design interventions.

1. INTRODUCTION

Ship designers were always motivated to reduce power demand, but in past the main aim was to reduce costs. Throughout last decade International Maritime Organization (IMO) introduced Energy Efficiency Design Index (EEDI) starting operatively from 2013. and lately Energy Efficiency Existing Ship Index (EEXI) which will start with application from 2023. Together with the global imperative of reduction of greenhouse gas emissions and high prices of energy sources, strong need was created to continuously investigate all aspects which could contribute to reduction of power demand for ships.

Energy Efficiency Design Index (EEDI) is technical measure which aims at promoting the use of more energy efficient equipment and engines[1]. New ship design needs to meet the reference level of this index since 2013. The level to be achieved is incrementally changed every five years, so continuous work on innovations and development is fostered. This regulatory regime incorporates about 85% of CO2 emissions from international shipping[1]. Energy Efficiency Existing Index (EEXI) is a measure introduced by IMO lately and first periodical survey is set for 2023[2]. This measure encompasses even larger number of ships since it concerns all ships of certain types which are currently in service. Both indexations are serving International Maritime Organization goal to reduce greenhouse gas emissions by 40 percent until 2030 and subsequently another 50 percent until 2050[2].

Upon introduction of EEDI it was quickly recognized that for Ro-Ro cargo and Ro-Ro passenger ships having relatively high installed power, variety of conditions in operational profile and service speed which tends to be as high as possible, there is large scatter in EEDI values. Therefore, there are guidelines and recommendations how to apply for correction factors to get proper EEDI reference line[3]. These types of ships were even excluded from the implementation of the required EEDI in the first phase until 2014. Furthermore, many of Ro-Ro passenger ships are intended for domestic routes solely and in such cases EEDI indexation is not applied, since for the ships solely engaged in voyages in waters subject to the sovereignty of the State flag indexations is not applied[1]. So, design of Ro-Ro ship presents special challenge with respect to energy efficiency and even if it is not intended for longer journeys, existing regulations and guidelines are of significant help throughout design process. In short term such ships will also benefit from introduction of energy efficiency measures, and long-term it can be expected that the regulations will get stricter in the future and probably encompass wider range of ships.

Design measures related to energy efficiecny for ships include as follows: modification of hull parameters, optimization of propulsion systems, hybrid propulsion systems, alternative fuel sources (solar, wind, hydrogen fuel). Operational measures mainly include slow steaming, route optimization and trim optimization[4].

Within this paper focus was to conduct hydrodynamic research for the projectof Ro-Ro passenger ship considering that from the start the project was very carefully planned and already encompassed specific measures to eliminate emissions and reduce consumption. The project of 105 m Ro-Ro passenger ship is designed to be fully electrically driven. With such approach emissions are already eliminated, but any possibility of saving related to energy consumption is highly appreciated because of the capacity of batteries and time needed to charge them. Regularly, technologies related to the electric propulsion on-board are as follows: diesel-electric drive (diesel generators generate electricity which drives the electric engine), hybrid drive (batteries are on board in addition to the internal combustion engine) and finally fully electric drive where there is no internal combustion engine and only energy source are batteries[4]. Such technologies are suitable for ships which sail shorter distances and the project of concern intended for routes in Adriatic Sea is good example of applicability of electric drive. When it comes to the efficiency of the batteries throughout last decade significant advance is achieved and this enabled transferring technology from small and slow ships to projects of larger and faster ships.

Starting point for the energy efficiency analysis of the ship is the quality of the hull itself i.e. resistance of the ship and propulsive efficiency. Findings which are presented in this paper are actually simple interventions applicable for both, newbuilds and the existing hulls. Ship hull with optimized hull lines meaning that for the selected main dimensions lowest resistance was already achieved, was investigated with respect to appendages resistance and choice of propeller. Set of towing tank test was conducted and modifications applied were not demanding for preparation. However, indicative, and measurable gain is achieved.

2. HYDRODYNAMIC ASPECTS FOR RO-ROPASSENGER SHIP WITH PODDED PROPULSION

Hydrodynamic aspects of such project encompass wide range of project parameters and characteristics. Focus of this research is to analyse ship resistance and propulsion efficiency. Other items of interest may be propeller hull excitations, erosion, cavitation of propeller and appendages, seakeeping qualities (slamming, motions, green water, swinging), manoeuvrability (course stability etc.)[5].
Usually, for the initial phase of the project it is helpful to rely on rules of thumb for different hydrodynamic aspects[7]. This particular project already had optimized hull lines with the aid of numerical analysis and further modification of hull dimensions or lines was not considered. Effort was put in systematic towing tank testing of certain variations of the appendages and propeller, which proved to be cost-effective.
Increase of total resistance due to appendages for ships with classic propulsion which include rudders, shafts, bossing, struts, fin stabilizers etc. can easily reach over 10%. For ships with podded propulsion the appendages are usually limited to side thrusters and headboxes, so their total resistance is much less compared to classic propulsion. According to literature [5] for well streamlined headboxes the resistance can be reduced to as low as 0.5%, or in case of not so good solutions this value can increase to more than 2%. Resistance of stern thrusters is usually negligible, and as to the bow thrusters it can reach 3-4% percent depending on the number, size, and location[5].
Pod housings also create resistance, but they are considered as a part of propulsive system and not included in the total ship resistance. Some studies indicate that solely the resistance of housings can reach 4% per one housing, depending on the size and design [6]. In analysis, influence of housing is included in propulsive efficiency.
Generally, resistance of the aft body of ship is mainly viscous resistance (friction and pressure resistance)[5]. Reduction of frictional resistance for a given speed is possible with increased Reynolds number (length). Pressure resistance can be reduced by reduction of loss of static head to a minimum which actually implies reduction in momentum losses (flow acceleration and separation) in the boundary layer [3].When the transom stern becomes immersed, this leads to the separation as the flow under the transom passes out astern of the hull. Consequently, vorticity occurs in the separated flow behind the hull and pressure loss is inevitable. Drag which occurs due to the transom immersion is largely a pressure resistance and it is scale independent [6].
Additionally, stern devices such as wedges, interceptors or duck tails are also considered as appendages and their purpose is to reduce aft body resistance. For smaller vessels, elements such as interceptors or wedges primarily improve dynamic trim with respect to power demand. For larger vessels, such elements serve for smoothening of stern wave, and they decrease the pressure drag due to flow field change [9]. For instance, with the interceptor or stern wedge local pressure increases with the deadwater area created in front of it. Flow over the aftbody of a ship bends downward creating lift. Most drag reduction is due to the pressure drag reduction including wave making resistance [9]. Duck tail application should lead to the reduction of the stern wave height and improved propulsion efficiency. For vessels sailing at Froude numbers 0.23 – 0.3 it is recommended to increase the length of the aft body [5]. In such way,it is expected to reduce height of the transom wave pattern and improve trim of the vessel [3]. Based on literature, combination of duck tail and interceptor can save 3 – 12 % in power and the effects are significantly varying over the speed range [3].
Bow thrusters are for passenger ships often inevitable. Since the location of the thruster as well as its diameter are mainly determined by the manoeuvrability demands, possible interventions mainly include scallop and grids. As it is known, tunnel step in the rear is decreased by application of scallop, additional form resistance should be decreased[7]. But some investigations concluded that application of scallop may even cause resistance increase and recommend for the scallop to be carefully chosen by investigating the characteristics of local flow around the tunnel inlet[9]. By adding grids to the thruster's tunnel it is possible to reduce opening resistance by 1% per opening [5].
Approach applied within this investigation included analysis of duck tail, wedge, and interceptor's configuration and additionally application of grids for bow thruster. Besides encompassing such analysis in design phase, it is fully applicable for existing ships as well. Following the results of this and other published research related to appendage drag, it is recommendable to conduct survey for the ships in service and if needed undertake measures concerning appendage redesign, since lot of such work is fully feasible even after the ship is delivered.
Towing tank tests are suitable means to verify design interventions and as it is explained, pressure drag should not be affected by scale. However, differences in thickness of boundary layer in the aft body of the model and full-scale ship may affect obtained results and should be considered.

Second stage of investigation concerned propulsion, specifically propellers. Ships with podded propulsion feature better wake fields, especially in the case of pull type propulsor which is of course beneficial for the propeller design. Other advantages are more freedom in locations of propulsion units, low appendage resistance and no need for stern thrusters [5].

3. MODEL TEST DESCRIPTION

Model tests were conducted in the large towing tank of Brodarski Institute – Croatian Shipbuilding Corporation in Zagreb, Croatia. Large towing tank is 276 m long, 12.5 m wide and 6 m deep. The model scale was chosen as small as possible considering the tank size. In such manner it was achieved that for the range of Froude numbers of interest Fr = 0.2 – 0.3, Reynolds number for the model exceeds $Re=1\times10^7$. In such manner, it is to expect that potential scale effect should be less pronounced which makes testing more reliable.
Since very small changes were expected with the application of certain modifications, particular care was taken related to repeatability. Measurement signals were extraordinary long and basic resistance test of the hull with initial configuration of appendages was repeated three times throughout the test sequence.
During resistance tests following values were measured: model total resistance RT, dynamic trim angle and rise/fall at LCB. Since visual observation of potential changes to the stern wave is not easy due to minor changes included, extra measurement of the wave height at the stern is suggestable. Another phenomenon that cannot be observed during model test is concerning vorticity effects and this is where CFD simulation represents complementary analysis. Anyhow, the extent of data obtained through these model tests was enough to demonstrate differences and enable decisions related to proposed solutions.
Second stage of investigation included self-propulsion tests and these tests were conducted in line with ITTC recommendations. Thrust, torque and revolutions per unit were measured and analysed together with the results of resistance tests and open water tests of the units and propellers alone. Housings of the pods were considered as part of propulsive system i.e. not included in the appendages of the model during measurements[8].

Figure 1: Ro-Ro Passenger ship model during self propulsion test

4. RESISTANCE TESTS OF APPENDED HULL

Since the hull with podded propulsion features less appendages compared to the classic propeller propulsion, possible interventions related to the appendage resistance can include side thrusters and stern devices such as duck tail, wedges, or interceptors. The room for improvement is limited, but viable solutions are still applicable.

4.1 INVESTIGATIONS OF DUCK TAIL VARIANTS

For this design it was decided in advance to apply duck tail since it was proven to be successful for some previous designs. So, basic influence of the initial configuration of the hull with and without duck tail was not investigated except for qualitative overview gained through numerical simulation, which has showed positive impact.

Investigation included changing of the angle of the duck tail towards the water line i.e. effectively creating a wedge, adding interceptors to the duck tail and elongation of the duck tail. Iterative resistance tests were conducted in order to investigate possible stern variants and altogether seven different configurations were tested (labelled C1 – C7).

First set of tests involved variations of the angle of duck tail towards the waterline creating actually the wedge. Length of the duck tail was not changed. Initial configuration C1 and two variations C2 and C3 were tested (shown in Figure 2 below). Influence upon percentage of change in total model resistance is shown in Figure 3.

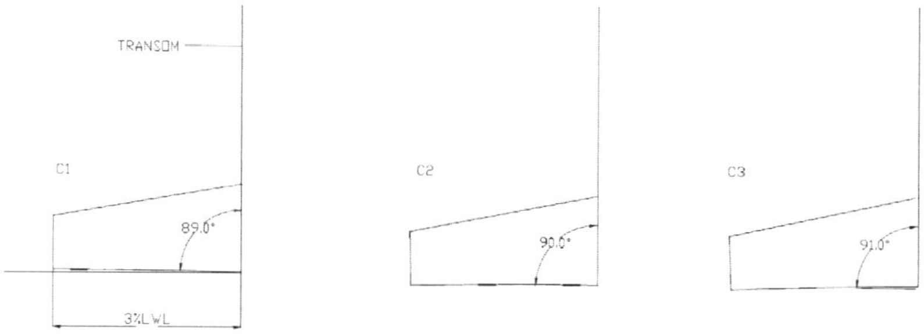

Figure 2: Variations of Duck Tail angle towards transom: C1, C2 and C3

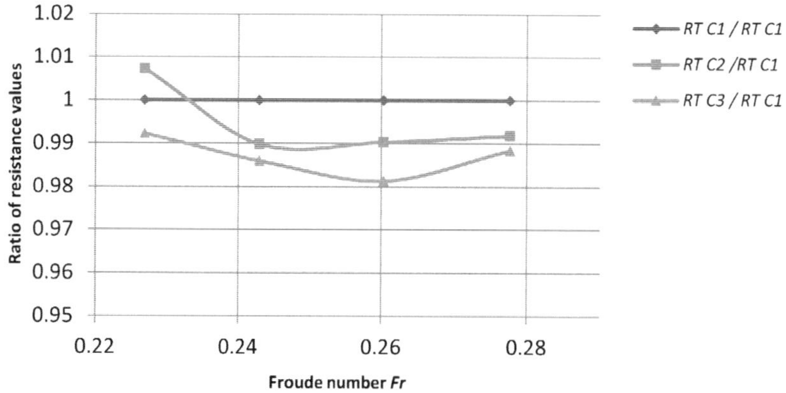

Figure 3: Ratio of the total model resistance*RT* of setting C2 and C3 related to setting C1

Setting C2 shows some improvement compared to C1 and for setting C3 this improvement is more pronounced. With C3 setting, improvement of 2% of total resistance is achieved compared to C1 for Froude number 0.26. Positive impact is present for this setting for the entire tested range of Froude numbers i.e. *Fr*=0.225 – 0.275 and the average reduction in total model resistance*RT* was 1.3%. Rise / fall values at LCB were changed insignificantly (less than 1 mm in model scale), while dynamic trim values were changed as expected. With the initial configuration dynamic trim by the bow was present (maximal value was less than 0.3 deg.). With duck tail settings C2 and C3 trim by the bow values were increased up to 25%.

Next step consisted of adding the interceptors to the configuration C3 which was best among three initially tested. Interceptor was aligned along the duck tail (shown in Fig 4). Configuration C4 contained interceptor height 0.03% L_{WL} (length at the water line) and C5 interceptor height 0.06% L_{WL}. Figure 5 shows impact of the application of interceptor added to the duck tail on total resistance of the hull.

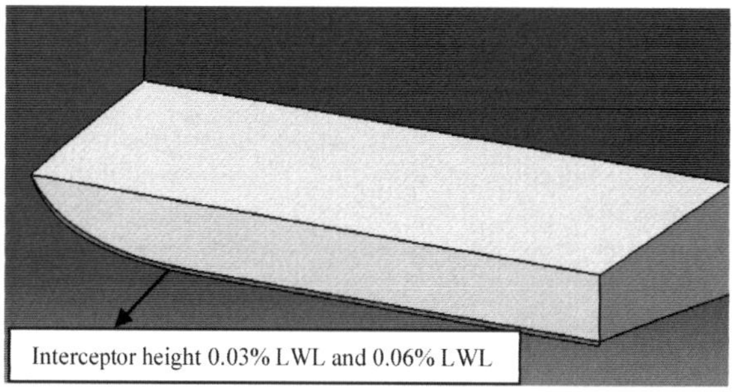

Figure 4: Geometry of interecptor added to the duck tail setting C3: C4 and C5

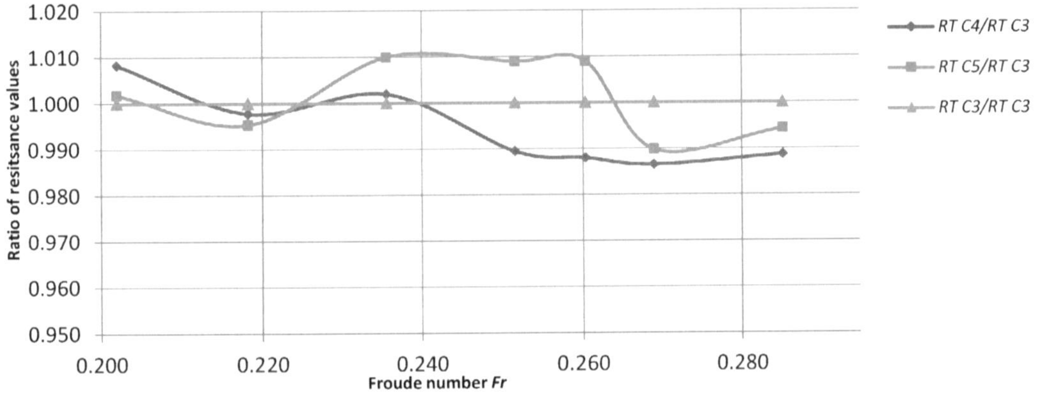

Figure 5: Ratio of the total model resistance*RT* of setting C4 and C5 related to setting C3

With the setting C4 further reduction of approximately 1% of total model resistance *RT* is gained in the range of Froude numbers Fr = 0.25 – 0.285, while for the setting C5 for Froude numbers Fr=0.23 – 0.26, deterioration of approximately 1% in total model resistance *RT* is present. For Froude numbers above Fr=0.27 the gain is similar for both configurations, C4 and C5. Changes in the dynamic trim angle were for C4 and C5 configuration mild. With such short investigation, actual optimized point for interceptor was found. Maximal gain is obtained with setting C4 which yields 3% of total model resistance *RT* improvement compared to setting C1 for the Froude number Fr = 0.26.
Third step of investigation included attempt to achieve similar effect as the one gained with interceptor, but with prolongation of the duck tail. Two additional settings were considered, C6 which includes prolongation of 0.45% L_{WL} and C7 which includes prolongation of 0.9% L_{WL}. These values were compared with the configuration C1 i.e. initial setting.

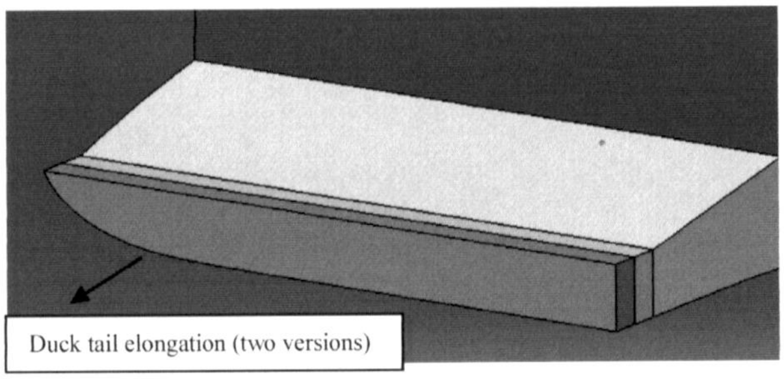

Figure 6: Overview of prolongation of duck tail (in cian and magenta): C6 and C7

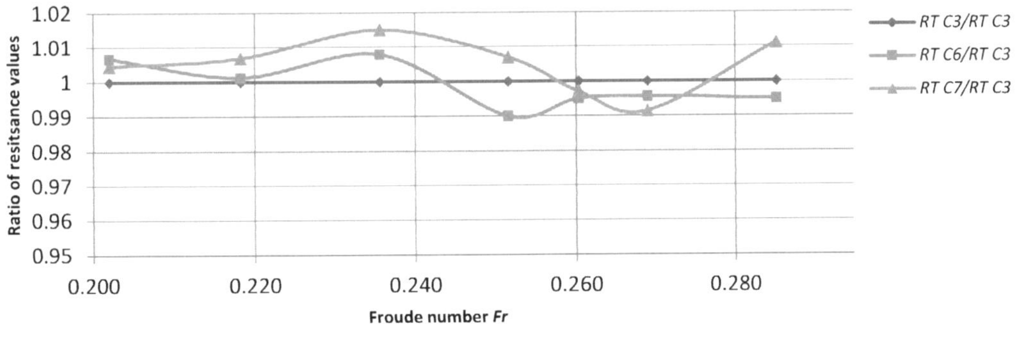

Figure 7: Ratio of the total model resistance *RT*of settings C6 and C7 related to settingC3

Improvement is present for Froude numbers *Fr*>0.25 for configuration C6. Gain is somewhat lower compared to the gain with interceptors and one can see from the graph that the measurement had more scatter and therefore it can be considered less reliable. One of the reasons is that when the differences in measurement become less than 1% it is exceedingly difficult or even not possible to prove such minor differences with the measurement, and other reason for scatter is most probably associated with vorticity which causes non-stationary phenomena. To sum up, for Froude numbers *Fr*=0.2 – 0.25 the total model resistance *RT* of the hull actually increases for up to 1.5% by prolonging the duck tail, while for *Fr* = 0.25 – 0.285 the prolongation yields small improvement, 1% or less. Such solution seems to be not beneficial for the entire operative speed range and therefore not so favourable for application.
Conclusively, compared to the initial setting best improvement is achieved with setting C4 comprising of duck tail with wedge formed with extra interceptor of height of 0.03% L_{WL}. Maximal gain measured is for *Fr* = 0.26 and it amounts 3% of total model resistance *RT*. Average improvement in total resistance of setting C4 for the range of Froude numbers *Fr* = 0.22 – 0.28 is 1.8% of total model resistance *RT* compared to setting C1.
Conducted tests clearly indicate limits of possible gains through application of stern devices for this project, i.e. one can say that possibilities of power reduction through intervention at the stern are fully explored, and the achieved savings are satisfactory. Interventions which are proposed are of simple geometry and not demanding related to technological possibilities, which makes them easily applicable for existing ships as well.

4.2 INVESTIGATION OF BOW THRUSTER GRID

The only other existing appendage for this passenger ship with podded propulsion is side thruster. For this project streamlines test by means of paint was conducted first and based on the results, from the bow side, flow inlet was shaped in a way that it was aligned with the streamlines test results. Since for some previous projects the positive effect of such approach was already proven, it was not evaluated separately. Fourth step of investigation was to install grid at the bow thruster opening and the total model resistance *RT* of the hull was reduced for 1.4% in average for the Froude numbers *Fr*=0.25-0.29, while for *Fr*=0.2 – 0.24 the influence was negligible. Such result is in line with the data from literature [4].

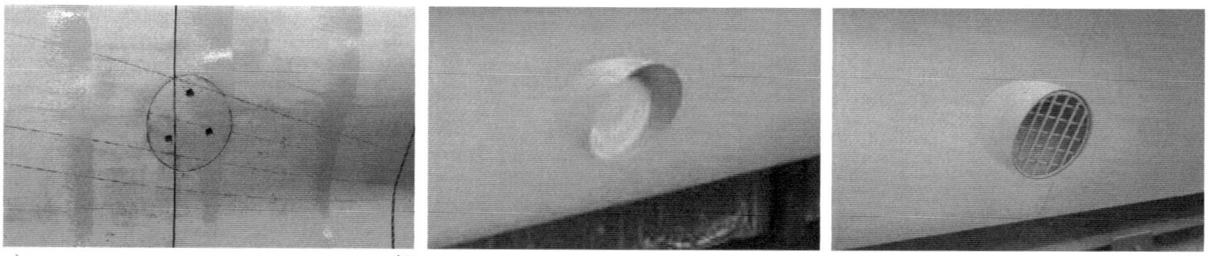

a) b) c)
Figure 8: a) streamlines test b) shaped inflow area c) grid installed at bow thruster

5. SELF PROPULSION TEST RESULTS

Self-propulsion tests with the initial stock propeller and housing configuration gained promising results, but since this first configuration was chosen somewhat conservatively with respect to calculation of cavitation margin and the propeller which was most suitable regarding characteristics was actually CPP, it was decided to conduct further research. It was of outmost importance to recognize all the potentials for saving and decision was to conduct second iteration of self-propulsion testing. Analytic approach was applied based on investigation of influence of main propeller particulars upon propulsive coefficients and consequently, total propulsive efficiency η_D.

Iterative analysis included alteration of the propeller number of blades, pitch, and area ratio of the propeller. Based on the obtained results, WB series propeller was chosen as next stock propeller. Since the hub of this propeller was smaller, pod housing was also slightly adapted geometrically to the new hub, so the new housing was slightly smaller. Propeller diameter was not changed, and the main parameters are presented in the following Table.

	Final Stock Propeller Prop No.2 (Version V2)	Initial Stock Propeller Prop No.1 (Version V1)
Pitch ratio at 0.7R ($P_{0.7R}/D$)	1.16	1.28
Area Ratio (A_E/A_0)	0.40	0.48
Hub Ratio (d/D)	0.18	0.26
Number of blades (Z)	3	4
Type of propeller	FPP	CPP

Table 1: Main characteristics of two stock propellers

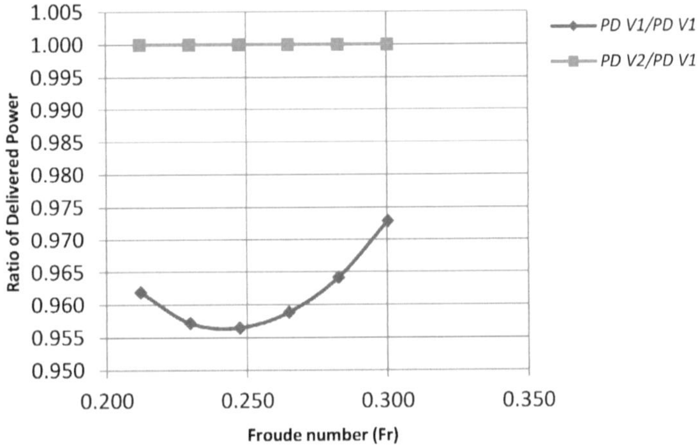

Figure 9: Ratio of the ship delivered power for two tested solutions

Repeated self-propulsion test for design draught revealed the improvement in total propulsive efficiency η_D of 3.8% in average for the range of Froude numbers Fr=0.21-0.3 which is presented in the Figure 9. Gained improvement indicates that for the ships in operation is recommendable to consider retrofitting. It is useful to check wake field and margin towards cavitation through tests or simulation.

6. CONCLUSIONS

This paper presents simple hydrodynamic interventions on the 105 m Ro-Ro passenger ship project comprising of changes related to geometry of appendages and analysis of propeller characteristics. Project investigated is podded propulsion ship which means that the setting of the appendages comprising only of side thrusters and some stern devices yields in start much smaller resistance compared to set of rudders, shafts, struts etc. characteristic for standard propulsion. The particular project is designed to be electrically driven and operate in domestic waters, so EEDI Indexation is not to be included in this project. However, recommendations for energy saving have to be checked and adopted maybe even more carefully, since the power demand on batteries should be as small as possible.

Results gained through model testing indicated maximal saving of 4.5% related to model total resistance RT decrease (changes in duck tail setting and adding bow thruster's grids) and also maximal saving of over 4% is achieved related to the reduction of ship delivered power PD in self propulsion condition. For the particular project when recalculating the measured model scale saving of 4.5% to full scale it amounts actually 6.6% less effective power P_E. In average in the wider tested range of Froude numbers Fr=0.22 – 0.28, saving in total model resistance was 1.8% which means 3% less in effective power P_E for ship.

Self-propulsion tests were conducted with the same appendage setting (C3) and achieved saving was 3.8% with respect to lower power demand i.e. higher efficiency achieved with new propeller for the range of Froude numbers Fr = 0.22 – 0.28. Therefore it is not possible to define exactly cumulative gain i.e. total reduction in power demand which would be achieved with the best configuration regarding resistance and better option of stock propeller, but it can be estimated that it amounts over 10% for Froude number Fr=0.26, if the propulsive coefficients related to the hull-propeller interaction are not deteriorated.

It is important to emphasize that hull lines were developed on the basis of vast design experience accumulated by designer on previous projects and then also optimized with the aid of numerical methods. Overall saving is relevant, and the reduced power requirement is to be beneficial.

Presented interventions are easily applicable and can be quickly tested and verified, enabling cost-effective improvements. Viable ways to check in advance the benefits of such interventions would be to conduct model tests, potentially together with numerical simulations for more detailed approach. Related to the stern devices such as duck tail, wedges, or interceptors, for future investigations it would be beneficial to measure wave elevations behind the stern. Visual observation is not sufficient for such minor changes in final result. Numerical simulations mentioned could give more info related to vorticity phenomena and pressure change. However, such detailed approach would be useful for research work and for operative purposes similar set of model tests is sufficient to make further decisions.
Investigation also clearly indicated limits of improvement related to applied devices i.e. for the particular project no further gain can be expected in relation to the appendages setting. Improvement gained in self propulsion condition is informative for the propeller designer and also refers to the potential of reconsidering retrofitting of the propeller for ships in operation.
To conclude, nowadays in respect to demands on greenhouse gases reduction together with extremely high prices of energy sources, it is even more important to explore all the possibilities of savings and provide energy efficient solutions. Approach and interventions presented in this paper are simple for application and quick for evaluation. Such tangible results obtained for project with optimized hull lines are encouraging for shipowners and designers to research to investigate every aspect of hydrodynamics related to energy saving.

7. ACKNOWLEDGEMENTS

Author expresses very great appreciation to Design company CLASSIS and Shipping company JADROPLOV from Split, Croatia for making this research feasible and showing high level of engagement for progress related to energy saving. Special thanks to Ship Designer Predrag Čudina, PhD for cooperation and support.

8. REFERENCES

1. Imo. Technical and operational measures, n.d., (online) Available at https://www.imo.org/en/Ourwork/Environment/Pages/Technical and Operational Measures.aspx

2. Available at; https://www.imo.org/en/MediaCentre/HotTopics/Pages/Cutting-GHG-emissions.aspx

3. ALISAFAKI, PAPANIKOLAU A.D., 'On the energy efficiency design index of RO-RO passenger and RO-RO cargo ships,' *Proceedings of the Institution of Mechanical Engineers, Part M: Journal of Engineering for the Maritime Environment*, 231 (1), 2017.

4. BARREIRO, J., ZARAGOZA S., DIAZ-CASAS V.,'Review of Ship Energy Efficiency', *Ocean Engineering 257,* 2022.

5. HAMALAINEN R., HEERD J., 'Energy saving Possibilities in Twin or Triple Propeller Cruise Liners', *3rd International Symposium on Marine Propulsors, Launceston, Tasmania, Australia*, 2013.

6. HAMALAINEN R., HEERD J., 'Hydrodynamics development for a large fast monohull passenger ferry', *The Society of Naval Architects and Marine Engineers*, 1998.

7. CARLTON J., 'Marine Propellers and Propulsion', *Elsevier*, 1994.

8. ITTC Procedure 'Recommended Podded Propulsion Test and Extrapolation', available at https://www.ittc.info/media/8015/75-02-03-013.pdf, 2017.

9. JANG H.S., LEE H.J., JOO Y.R., KIM J.J., CHUN H.H. 'Some practical design aspects of appendages for passenger vessels, *'International Journal of Naval Architecture and Ocean Engineering,* 2009.

9. AUTHORS BIOGRAPHY

Marta Pedišić Buča is Head of Hydrodynamics Department at Croatian Shipbuilding Corporation / Brodarski Institute, Zagreb, Croatia. She has more than 20 years of working experience in hydrodynamic facilities and is responsible for towing test organization and conduction. Previous experience includes numerous projects of hydrodynamic testing, including at least hundred projects which encompassed some kind of optimization for the purpose of energy saving. Besides main engagement, she was teaching at Faculty of Mechanical Engineering and Shipbuilding, University in Zagreb for 15 years and is actively involved in AC activities of ITTC, as Chair of one Working Group.

THE ROLE OF GRAPHENE IN DECARBONIZATION. THE STOLT TANKERS EXPERIENCE.

J A González and G Zommers, Stolt Tankers, The Netherlands

SUMMARY

Graphene consists of monolayers of carbon atoms bonded in a repeating hexagonal pattern, it is the thinnest known material, and it is very well known for its thermal, conductivity and lack of biota adhesion properties. It was isolated in 2004 at the University of Manchester by Professor Andre Geim and Professor Konstantin Novoselov, who went on to win a Nobel prize for their discovery. Stolt Tankers is committed to the reduction of the environmental impact of our fleet activities and is involved in the transition to GHG reduction in 2030 and 2050 whilst keeping our goals. The negative effects of biofouling adherence to the surfaces in contact with the seawater are well known, its impact is becoming more visible since new IMO regulations will be in place (EEXI, CII). Stolt Tankers pioneered the application of graphene coating in a propeller of a deep sea-going chemical parcel tanker in April 2022; it represents the first time that it has been done aboard a tanker worldwide. The present paper will provide some of the project's initial findings and the impact evaluated so far. There are plenty of areas where graphene applications could help the marine industry in the energy transition is embarked, the paper will present some of the options available.

NOMENCLATURE

CII	Carbon Index Indicator
DWT	Dead Weight Tonnage
EEXI	Energy Efficiency Existing Index
GMTT	Global Marine Technology Trends
GRT	Gross Tonnage
IMO	International Maritime Organization
nm	nano meters
MCR	Maximum Continuous Rating

1. INTRODUCTION

Biofouling is a major contributor for energy efficiency reduction, the ships spend almost their entire operational life afloat favoring the creation of biological layers (biofilm) and macrofouling populations in different working areas of the ship. The consequence of this occurrence is mainly reflected in additional dragging, losses of thermal performance of facilities, economic losses, inability to carry out proper maintenance practices, etc. It is important to recognize that biofouling is an important issue, resulting in significant economic, environmental, and technological impacts.

The new regulations and decarbonization goals promulged by IMO are forcing all marine stakeholders to find suitable solution to decrease vessels' environmental footprint. There is a wide range of options that can be tested/applied; this paper presents the early results of the experience gained with one of them i.e., application of graphene coating to a propeller of a ship aiming to decrease the fouling rate. Graphene is a relatively new material due it was isolated in 2004; its potential for application in the shipping industry was not studied and highlighted until the launch of the Global Marine Technology Trends 2030 report GMTT 2030 (Lloyds Register, 2015) [1]. In that piece it was identified the exceptional combination of strength and toughness of metals that can be achieved by nanoprecipitation, such as doping nano sized carbide or copper precipitates; and the graphene's anti-bacterial properties will create new opportunities for application in fields like hull coating. By means of graphene films, hull surfaces will be covered, avoiding micro- and macro-fouling attachments. The robustness and resistance of graphene are useful characteristics for this type of application.

2. BIOFOULING AND DECARBONIZATION

Fouling refers to the formation of deposits on the equipment surfaces. This type of deposition originates significantly decreases on the equipment performance and/or the operational lifetime of the equipment. One of the fouling types is the biological fouling -biofouling- (Epstein, 1981) [2], it refers to the settlement and growth of microorganisms (biofilm) and macroorganisms such as mussels, oysters, barnacles, etc.

Biofouling is the representation of a multi-faceted issued that heavily impacts the floating assets, its origin is purely biological since under natural conditions, 90% of microorganisms in the biosphere live in biofilms; bacteria attach to conditioned surfaces and form a slime layer. The advantages of being attached to a surface include the continuous supply of nutrients, protection and the removal of metabolic waste products. The continuous movement of the aqueous environment over a surface also usually means that the water is aerated, which is an advantage for aerobic bacteria (Costerton et al., 1995) [3]. The creation of complex structures called extracellular polymeric substances (EPS) in which the microorganisms are embedded, facilitates the development of specific microbial communities (Characklis & Wilderer, 1989) [4].

Biofouling is a well-known problem in the marine industry. The hull of the ship is constantly colonized both by micro and macrofouling in several different locations. Some of them could be rudder or seachest. This fouling implies great losses linked to among others, the vessel's speed, fuel consumption increase, drag, less autonomy, the increase of the cleaning operations and maintenance of equipment and associated docking time (González et al., 2012) [5]. The US Navy carried out a study about hull fouling economic impact on a medium size combat ship Arleigh-Burke class destroyer (Schultz et al., 2010) [6], it found severe associated costs due to the hull fouling hull included those related to extra fuel consumption and to the hull cleaning routines. The losses due to the hull cleaning and reapplication of coating are fewer than the ones due to extra fuel consumption. An estimation of all these expenses is 56 million of $ per year for all the destroyers in the Arleigh-Burke Class. The decrease of heat transfer efficiency is another example of biofouling impact; it has been identified that the impact of microfouling is causing losses linked to the OPEX of the heat exchanger-condensers and consequently increasing the fuel bill and emissions footprint. The mitigation of microfouling in heat exchangers used in marine applications will significantly improve the efficiency of the whole propulsion package installed on board (González., 2019) [7]. Biofouling has unveiled itself as one of the areas with major potential to reduce emissions of the fleet and reach the emissions targets for 2030 and 2050

3. GRAPHENE – IS A CARBON FORM ONE OF THE KEY ELEMENTS FOR DECARBONIZATION?

Graphene is an interesting material to be traced by the shipping community, the expectations with this material stem from its remarkable physical properties and the potential applications these properties offer for the future. Graphene's anti-bacterial properties might create new opportunities for application in fields like hull – propeller – niche areas coating. By means of graphene films, hull surfaces might be covered, avoiding micro- and macro-fouling attachments.

Carbon is one of the most important elements in our life, and it is the fourth most abundant chemical element in nature. Not only constitutes one of the basic elements for life (an average representation of carbon in the mass of living matter is 19.4%), but it is also widely used in industry for materials manufacturing. The fundamental feature of carbon is its unique capability for combining with other elements. For example, the so-called hydrocarbons are formed by the grouping of carbon and hydrogen atoms either in chains or in rings. The addition of methyl radicals, nitrogen, oxygen and other new elements provides more complex molecules (acids, alcohols, etc.). Graphene consists of a monoatomic layer composed of carbon atoms densely packed in a hexagonal lattice (2,500 billion at/cm2). Its strong bonds maintain the interatomic distance in 0.142 nm, the shorter distance of all the materials that provides to the graphene outstanding mechanical properties.

Although scientists knew one atom thick, two-dimensional crystal graphene existed, no-one had worked out how to extract it from graphite. In the early 20th century, some physics argued that strictly 2D crystals were thermodynamically unstable and could not exist (Landau., 1937) [8], (Peierls., 1935) [9]. In 1947, their theory pointed out that thermal fluctuation in low-dimensional crystal lattices should lead to displacements of atoms comparable to interatomic distances at any finite temperature. Yet in 1962, the word "graphene" started to be used in order to define a graphite monolayer. It was not until 2004 when the experimental discovery of graphene happened (Novoselov et al., 2007) [10]. Graphene was isolated in by two researchers at The University of Manchester, Professor Andre Geim and Professor Kostya Novoselov; they isolated for the first time a layer of graphene using ordinary scotch tape and graphite. This particular method, known as mechanical exfoliation, leaded to the acknowledgment in 2010 with the Nobel Prize in Physics for the discovery of graphene and its characterization.

From the main characteristics of graphene, several important properties can be derived; and some of them are, in addition, unique and extreme as: linear energy dispersion, standard sequence of the conductivity with the magnetic field observed in conventional Quantum Hall Effect (QHE) appears shifted in graphene. In the field-effect configuration, this implies that carriers can be tuned continuously between holes and electrons by supplying the requisite gate, this fact implies a wide range of mechanical, optical and thermal attributes, which are also relevant (Geim et al., 2007) [11]. This leads to some potential industrial fields of application of graphene as follows:

- Photovoltaics: in the shape of more efficient solar cells using graphene as a charge collector due to its energy absorbing characteristics.
- Conductive ink: in the shape of graphene-based coatings for printed electronics.

- Ultracapacitors and micro batteries: in the shape of devices able to store big quantities of energy (around 20 times more powerful than the existing ones and with a charging time capacity of 1,000 times faster).
- Biomedicine applications: in the shape of implants coated with graphene able to provide more mechanical resistance.
- Thermal management: high thermal conductivity and consequently fast cooling
- Graphene based quantum dots: nano light with extraordinary properties due to their remarkable quantum confinement and edge effects: systematic photoluminescent mechanisms, bandgap engineering, in addition to the potential applications in bioimaging, sensors, etc.

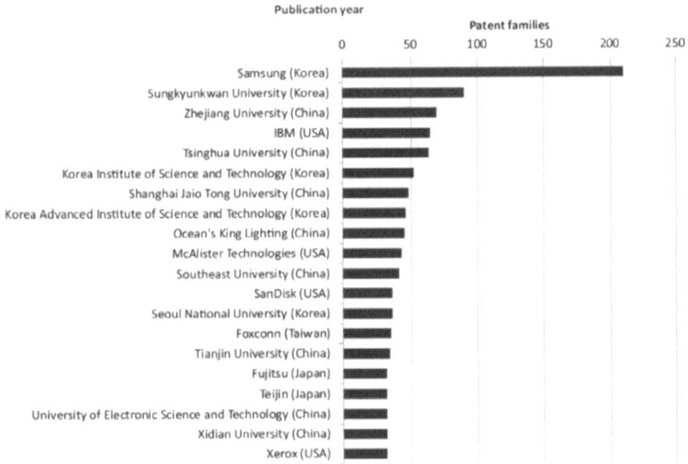

Figure 1: Number of patent families of the top 20 applicants (Intellectual Property Office, 2013) [12].

4. STOLT ACER

Stolt Acer is a chemical parcel tanker that belongs to Stolt Tankers fleet as part of its Asian fleet. Main characteristics of Stolt Acer are presented in *Table 1.* She was built in 2004 by Kitanihon Shipbuilding in Hachinohe, Japan. She has a total of 28 tanks able to allocate 28 different cargoes at the same time. She is powered by a 2-stroke engine able to deliver a total power of 7,980 kW at 120 rpm and equipped with 3 generators able to deliver a total power of 1,950 kWe.

Table 1: Stolt Acer main characteristics

Ship Name	*Stolt Acer*	***Length Overall (m)***	*170*
IMO Number	*9272668*	***Draught (m)***	*8.3*
Ship Type	*Chemical Parcel Tanker*	***Breadth (m)***	*25.63*
		DWT (Tons)	*29,709*
YOB	*2004*	***GRT (Tons)***	*18,703*

Figure 2: Stolt Acer (Photo credit: Stolt Tankers)

5. PROPELLER GRAPHENE COATING APPLICATION

Immediately before drydocking, the hull and propeller sequence of events was as follows; on the 14th of March 2021, a hull cleaning and propeller polishing service was carried out, it was followed by another hull cleaning and propeller polishing on the 17th of February 2022 which happened 2 months prior the ship was docked. The vessel arrived at the Gryfia Shipyard in Szczecin Poland for a scheduled Hull treatment drydocking. The Antifouling (AF) coating system applied at this docking was supplied by Jotun Paints and consisted of a combination of SeaQuantum Classic S (applied to the Vertical Sides) and SeaForce 60M (applied to the Flatbottom). The vessel was docked on the 6th of April 2022.

The diameter of Stolt Acer propeller is 5.5 meters having an area of 14.73 m2. The paint applied was XGIT PropPrimer (1 layer) plus XGIT Prop (1 layer), this is an ultra-low friction hard foul release system combined with a special primer that adheres well to bronze or similar alloys. Graphene plays an important role on creating the strength that increases resistance to cavitation and the low friction that provides propeller thrust efficiency. In preparation for the coating application, the ship's propeller shall be grit blasted to SSPC-SP-5 standards assuring the abrasive mixture will be of sufficient size to secure and anchor profile before receiving the new coating. Both layers were applied using spray method, when applying both layers for the first time, a trial must be conducted before spraying and all the humidity and external conditions must be checked to ensure a proper application and final thickness result implementation in the propeller; all surfaces must be clean and dry and oil grease and oil must be removed. Ship can be re-floated (or propeller submerged) 48 to 72 hours after final coating application, depending on outside environment temperature.

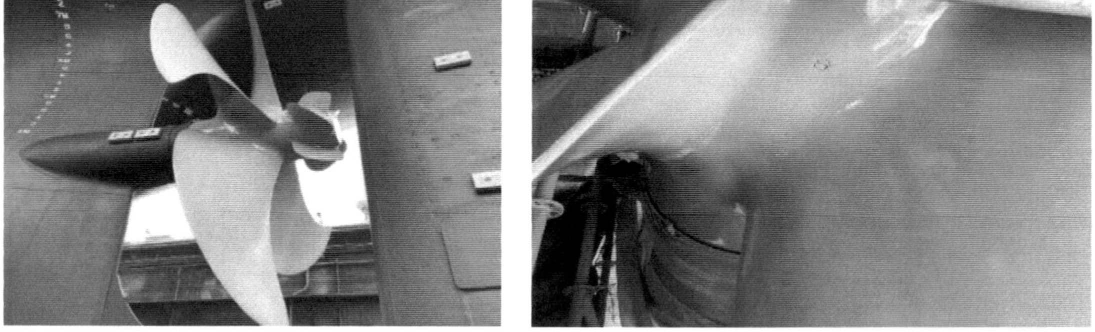

Figure 3: Graphene coating applied on Stolt Acer propeller

6. PRELIMINARY ANALYSIS

Data analysis has been carried out both by Stolt Tankers and coating manufacturer. The results presented in this paper are those carried out by Stolt Tankers. The analysis is based on noon report data received from Stolt Acer and comparison of consumption before and after application. The tool utilized for data visualization and result analysis is Storm Geo S-Insight. Data window for analysis purpose has ranged from 1st of January 2022 until 27th of September 2022; the blue marks correspond to trend before drydocking, the red marks correspond to the trend after drydocking. Filters applied have been draft 8.5 meters and above, main engine MCR from 40 to 100%, wind Beaufort up to 5, all data outliers have been filtered out and excluded.

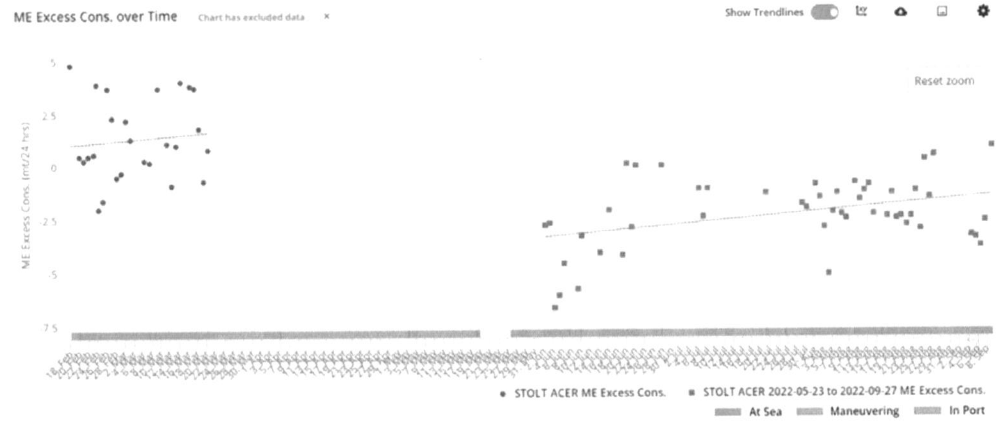

Figure 4: Main engine excess consumption over time dashboard

Before drydocking the excess consumption value for Stolt Acer was 1.33 metric tons of fuel per day, after drydocking as per the graph a value of -2.27 metric tons of fuel per day has been recorded. The difference between these two figures is 3.6 metric tons of fuel per day. When it comes to performance speed, before drydocking the value of performance speed for Stolt Acer was 11.97 knots, after drydocking the value recorded was 12.49 knots. The difference between these two figures is 0.52 knots. The difference before and after drydocking of the consumption performance in percentage is 20.3 %.

From above data it can be concluded that ship on average is consuming 20% or 3.6 metric tons per day less fuel than before drydocking for same speed. Taking into consideration a hull maintenance impact of circa 12-15% due to heavy fouling condition on arrival to drydocking, propeller maintenance impact is between 5-7% and what most important with no change in efficiency over the time. A potential saving of 3-4% due to graphene coating is confirmed by above data analysis.

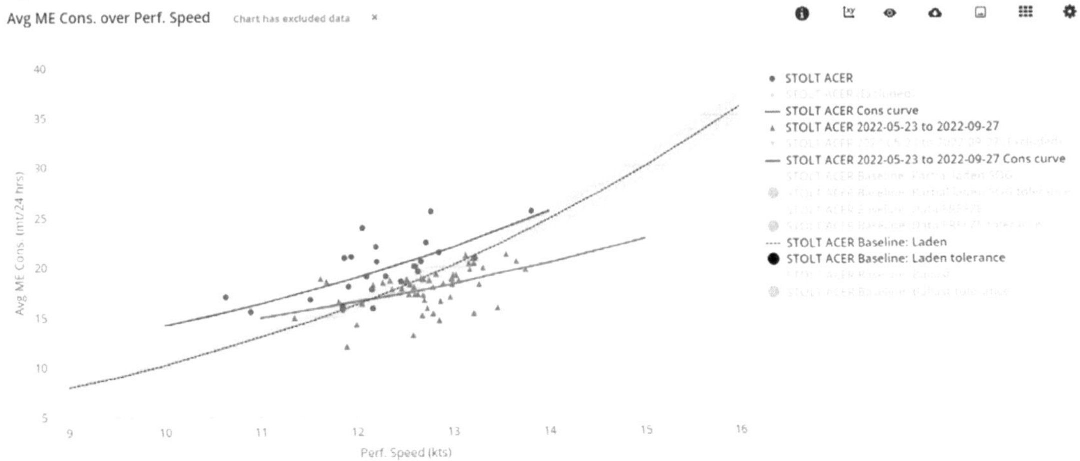

Figure 5: Average main engine consumption over performance speed dashboard

7. CONCLUSIONS

Biofouling is a worldwide phenomenon, arising in many different environments. Marine and industrial activities are especially prone to be affected by biofouling, as they are located directly in the marine environment. Biofouling can occur in many forms, but in general leads to major impacts in terms of increased OPEX and loss in operating profit. Graphene represents a very promising material for multiple applications not only in marine also in other industrial applications, for instance power generation, automotive and communications. It is envisaged that we will be more exposed to this material in the short term.

This article presents just one of the potential multiple applications graphene may have or shipping in the shape of an ultra-low friction coating applied to the propeller of a chemical tanker. This represents the first instance a chemical parcel tanker (Stolt Acer) has a propeller coated with this material. The results are very promising but not definitive when this article was submitted. Stolt Tankers has pioneered again a potential option for decarbonization and improving our emission ns and sustainability goals.

8. ACKNOWLEDGEMENTS

Stolt Acer crew; John Agustino (Stolt Acer Superintendent); Valentins Baburgs (Stolt Tankers Asia Fleet Manager); Maren Schroeder (Stolt Tankers Managing Director); Stolt Tankers Energy Conservation Team; Tobias Gröeger (Storm Geo); Marciel Gaier (GIT Inc.); Tim Kent (Lloyd's Register of Shipping); Professor Fai Cheng (Lloyd's Register of Shipping).

9. REFERENCES

1. LLOYD'S REGISTER OF SHIPPING., 'Global Marine Technology Trends 2030', *Report*, 2015.

2. EPSTEIN, N., 'Fouling technical aspects', *In: Somerscales EFC, Knudsen EJ (eds) Fouling of Heat Transfer Equipment. Hemisphere Publ. Corp., Washington, DC; 31-53,* 1981.

3. COSTERTON, J.W., GEESEY G.G., CHENG, G.K., 'How bacteria stick', *Scientific American*, 1995.

4. CHARACKLIS, W.G., WILDERER, P.A., 'Structure and function of biofilms', *Wiley Chichester Glossary*, 1989.

5. GONZALEZ, J.A., POLMAN, H.J.G., VENHUIS, L.C., BRUIJS, M.C.M., VAN AERSSEN, G., 'Biofouling issue, global implications and solutions linked to offshore business', *Marine and Offshore Renewable Energy: RINA Conference Proceedings*, 2012.

6. SCHULTZ, M.P., BENDICK, J.A., HOLM E.R., HERTEL, W.M., 'Economic impact of biofouling on a naval surface ship', *Biofouling*, 2010.

7. GONZALEZ, J., 'Application of quaternary ammonium compound and aliphatic amine as biocides in the mitigation of microfouling adhered to the internal surface of heat exchangers condensers refrigerated by seawater. Impact on the efficiency of different control strategies applied', *Taylor & Francis - Journal of Marine Engineering & Technology*, 2019.

8. LANDAU, L.D., 'Zur Theorie der Phasenumwandlungen II', *Phys. Z. Sowjetunion*, 1937.

9. PEIERLS, R.E., 'Quelques proprieties typiques des corpses solides', *Ann. I. H. Poincare*, 1935.

10. NOVOSELOV, K.S., JIANG, Z., ZHANG Y., MOROZOV, S.V., STORMER, H.L., ZEITLER, U., MAAN J.C., BOEBINGER, G.S., KIM, P., GEIM, A.K., 'Room-Temperature Quantum Hall Effect in Graphene', *Science*, 2007.

11. GEIM, A.K., NOVOSELOV, K.S., 'The rise of graphene', *Nature materials*, 2007.

12. INTELLECTUAL PROPERTY OFFICE., 'An Analysis of Worldwide Patent Fillings Relating to Graphene', *ipo.gov.uk/informatics*, 2013.

10. AUTHORS BIOGRAPHY

José González holds the current position of Energy Conservation Manager at Stolt Tankers. He is responsible for plan and manage energy projects, set up long-term vision and strategy on energy conservation field and execute those programs working together with other departments within the company. His career began as a seagoing engineer on board LPG gas tankers, bulk carriers and RoPax prior moving ashore. Over the last 15 years he has been working both in shipping and power generation businesses, leading Environmental, Sustainability and Energy departments for different companies like Westinghouse Nuclear Services, Cummins Engine Company, DNV GL Power Generation and Sustainability and Lloyd's Register of Shipping. Jose Gonzalez holds a Ph degree in microfouling impact on energy efficiency, he has a BSc in Naval Architecture degree and a MSc in Marine Engineering degree.

Guntars Zommers holds the current position of Energy Conservation Specialist at Stolt Tankers. He is responsible for plan and manage energy projects, assisting in set up long-term vision and strategy on energy conservation field and execute those programs on site by visiting ships, providing crew training, performing energy audits and providing ship performance data analysis reports. His career began as engine cadet onboard Stolt Tankers ship and he has been working in the Chief Engineer position before accepting a position in Energy Conservation Team within company. Over the last 8 years he has been working in Stolt Tankers Energy conservation department. Guntars Zommers holds a BSc degree in Marine Engineering.

ENERGY EFFICIENCY: DATA-DRIVEN APPROACHES TO HULL COATING SPECIFICATION AND SELECTION

R Mihaylova, **C Barnes**, Safinah Group, UK
M Porsbjerg, Safinah Group, Denmark

SUMMARY

This paper summarises some recent findings on selection trends and in-service performance of biofouling control coatings. The data on the arrival coating condition of ships was obtained via supervision of coating projects at dry dock. The paper also aims to highlight common issues with coating specification and selection that may result in sub-optimal in-service performance and provides recommendations on how to limit some of the most common issues observed while conducting systematic technical reviews of coating specifications as part of pre-planning activities.

1. INTRODUCTION

International shipping faces several challenges related to environmental performance, with the reduction of greenhouse gas emissions (GHG) being high on the international sustainability agenda.

In June 2021, the IMO adopted short-term measures to reduce GHG emissions as amendments to the International Convention for the Prevention of Pollution from Ships (MARPOL) Annex VI. The measures require ships to calculate the Energy Efficiency Existing Ship Index (EEXI) and to determine their annual operational carbon intensity indicator (CII). The CII rating scheme addresses operational performance by measuring how efficiently ships transport goods or passengers in terms of grams of carbon dioxide (CO_2) emitted against cargo carrying capacity and distance in nautical miles. From 2023 ships of 400 gross tonnage and above are required to have a ship-specific Ship Energy Efficiency Management Plan (SEEMP) onboard, which is required to outline how to improve the energy efficiency and reduce the carbon intensity through planning, implementation, monitoring, and self-evaluation and improvement. Ships of 5,000 GT and above of specific types engaged in international trade should have a verified SEEMP (SEEMP Part III), which documents the vessel's plans to achieve its CII targets. The intention is for continuous improvement, which will have to be achieved with a combination of various approaches and measures, including the use of energy efficiency technologies.

Hull coatings are usually claimed to offer between 1% to 10% GHG mitigation potential depending on type and operational assumptions [1], however, the *2021 IMO Guidance on treatment of innovative energy efficiency technologies for calculation and verification of the attained EEDI (Energy Efficiency Design Index) and EEXI* [2] states that 'category A' energy efficiency technologies, which includes hull coatings, affect the propulsion power and/or the reference speed of a vessel, and therefore their effect cannot be measured in isolation. Although the energy saving potential of a hull coating system and the effect on GHG emissions may be difficult to measure, poor in-service coating performance or the ability of the coating to control biofouling over time, can result in increased biofouling accumulation rates, which will have a direct impact on operational efficiency. Biofouling is the collective term used for the accumulation of aquatic organisms on surfaces and structures that are immersed in, or exposed to, the aquatic environment. The attachment of such organisms on ships' hulls translates into additional fuel costs and subsequently to increased emissions. A recent study commissioned by the GEF-UNDP-IMO (Global Environment Facility, United Nations Development Programme, and International Maritime Organization) GloFouling Partnerships Project collated findings on the impact of biofouling on GHG. According to the preliminary findings, a thin layer of slime (0.5mm) covering up to 50% of the surface can amount to 20% to 25% increase in GHG emissions, whereas a light layer of small calcareous growth could lead up to a 55% increase [3].

Biofouling has also been identified as a significant vector for the translocation of non-indigenous marine aquatic species, which could potentially become invasive. To address the risk of introducing potentially harmful species via biofouling, the IMO's Marine Environment Protection Committee (MEPC) adopted in 2011 the Guidelines for the control and management of ships' biofouling to minimize the transfer of invasive aquatic species ('the Guidelines') [4]. In 2018, MEPC 72 agreed to review the Guidelines; currently with a target date for completion of 2023.

Optimal performance in terms of biofouling control is associated with increased predictability / reduced risk of abnormal biofouling accumulation given specific conditions. A robust biofouling control strategy, ideally capable of coping with additional levels of uncertainty, should result in ensuring that suitable systems are selected and that they can perform as intended. The main components of such strategies are shown in Figure 1.

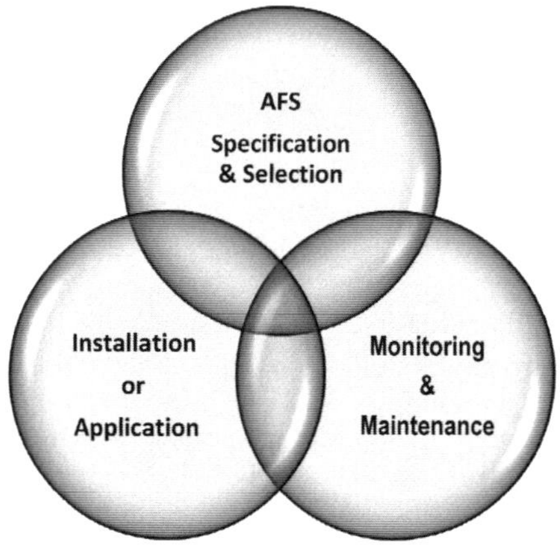

*AFS – means a coating, paint, surface treatment, surface, or device that is used on a ship to control or prevent attachment of unwanted organisms [4]

Figure 1: Main components of effective biofouling control strategies

This paper provides information on current hull coating specification and selection practices for the submerged areas of the hull, common issues that may lead to sub-optimal performance, and ways to address some of them by employing a systematic approach to this integral component of biofouling control strategies.

2. BIOFOULING CONTROL MARKET

For the purpose of this paper, 'coating technology' is defined as a family of coating products with similar mode of action and/or chemistry. For example, different types of fouling release coatings can be referred to as a 'technology'.

Selecting biofouling control for a specific ship is a complex task as there are multiple options available to choose from. These include manufacturer, coating technology, regional variations in formulations, biocide packages (if applicable), together with individual ship specific factors.

All of these choices affect the expected and actual in-service performance. Although environmental pressures encourage innovation, the increased market fragmentation, and the frequent introduction of new products with limited in-service track record, make the selection process even more complex (Figure 2).

Segment	Products	Total Products
Marine	378	512
Yacht	134	

*Multiple formulations under the same commercial name are included as different products. The list is not exhaustive.

Figure 2: Biofouling control market fragmentation

Despite the number of technologies and systems on the market, in-service coating performance data allowing independent comparisons between technologies / products and the effects of different operational conditions, has been limited. Data on coating condition assessments gathered during dry dock projects dating back to 2009 can be used to address this knowledge gap.

3. BIOFOULING CONTROL DATA INSIGHTS

A study based on a sample of 270 ships revealed that all ships had some level of biofouling present on arrival at dry dock [5]. According to the sample, 60% of the ships were reported to have less than 10% of the total submerged area of the hull covered in hard macrofouling, however, this can still result in significant fuel penalties depending on the biofouling composition and its dispersal across the hull. The data also showed that more than 40% of the ships were found to have in excess of 20% hard macrofouling across the flat bottom area.

The levels/extent of biofouling observed indicate that there is significant scope for improving performance at a vessel and fleet levels, including better consideration of operational parameters when specifying and selecting coatings.

4. CURRENT BIOFOULING CONTROL SELECTION PRACTICES

Selecting the optimal product and developing an appropriate final specification must be based on a careful consideration of a multitude of factors such as, but not limited to: operational profiles, activity, speed, environmental factors, and maintenance regimes. This is not just simply a process of opting for a product marketed as 'high performance'. Biofouling control products' perceived technology level, which is a function of marketing literature and product positioning, should not be treated as an objective indicator of in-service performance. Different products may be affected in different ways by specific operational or environmental conditions. For example, certain types of coatings, irrespective of perceived technology level, may be better suited to cope with the challenges posed by extended static periods in regions with high biofouling challenges.

In terms of coating selection, it would appear that about 70% of ships included in the sample had products perceived as 'high' technology level applied on the vertical sides at the last dry docking, as opposed to 43% at the previous coating application. This seems to indicate a move towards products marketed as offering high performance (Figure 3). However, it should be noted that the data includes a large proportion of ships representing the fleets of owners and operators who have pledged resources and effort to promoting sustainable shipping practices and may not be representative across all segments of the global fleet.

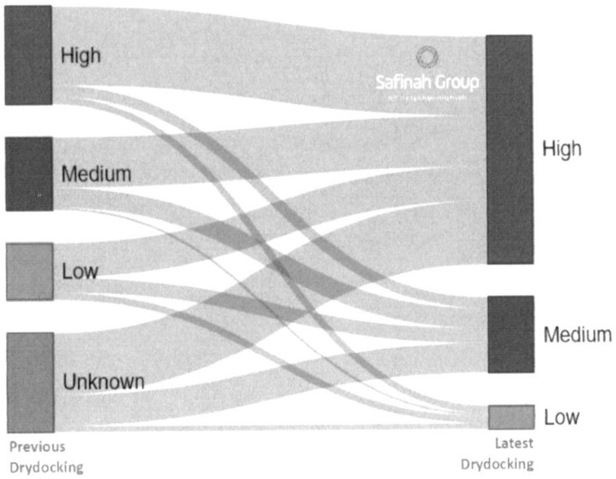

*Technology classification based on publicly available market positioning and other marketing literature.

Figure 3: Coating technology level transitions

Figure 4 illustrates the selection trends based on perceived technology levels and observed hard macrofouling on arrival at dry dock. More than half of the ships with less than 10% hard macrofouling on arrival changed manufacturer. However, half of the ships with more than 20% coverage remained with the same manufacturer, and often with the same technology or product applied at the next dry docking. This is a surprising finding, which could imply a commercially driven selection or a potentially long-standing relationship. It is also an indication that some owners may not carry out regular or adequately detailed fleet-wide reviews based on ship-specific performance to inform long-term strategy.

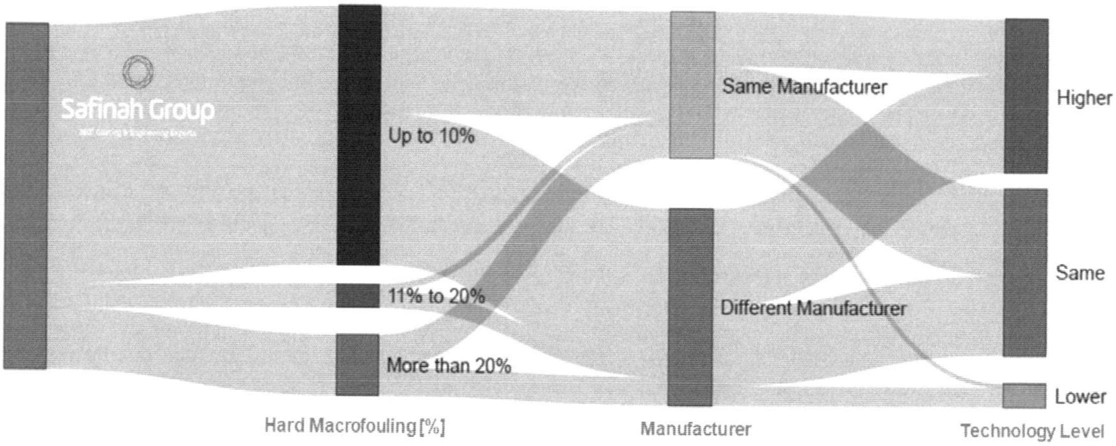

Note: The width of the flows represents the proportion of the sample corresponding to individual nodes within each vertical section. The sum of vertically aligned nodes represents the whole sample.

Figure 4: Selection trends based on observed hard macrofouling

5. COATING SPECIFICATION, SELECTION AND PERFORMANCE RELATIONSHIP

The apparent trend for choosing products marketed as offering superior performance is not surprising given recent market and regulatory drivers. However, applying advanced technologies does not necessarily result in superior performance, especially if technology attributes or specifications may not be tailored to the specific vessel.

Figure 5 provides a graphical summary of the operational history of a ship with a high technology level antifouling product that arrived in dry dock with more than 65% macrofouling on the vertical sides. Upon inspection it was determined that the coating on over 60% of the vertical sides was 'polished through' to the tie coat, which has no biofouling control properties. Such levels of biofouling accumulation can lead to a significant increase in fuel consumption. Premature 'polish through' (to tie coat) is a defect specific to biocidal antifouling coating systems, where the antifouling paint has 'polished away' exposing the underlying paint layer, often in lighter colour, referred to as 'tie coat'.

Topic	Highlights
Product	High level technology
Specification	• Days active [p.a.]: **50%** • Average speed: **14 kn** • Average SST: **25ºC**
Outcome	The specification did not match the historical or current vessel profile – direct impact on arrival condition

Factor	Pre-Drydock (2016-2018)	Post-Drydock (2018-2021)
Days Active [p.a.]	52% - 56%	52% - 56%
Speed >14kn*	24.8%	10.3%
SST above 25ºC	12%	28%
Static 14-21 days	NA	6
Static 22-30 days	1	3
IWC	Unknown	1 (2020)

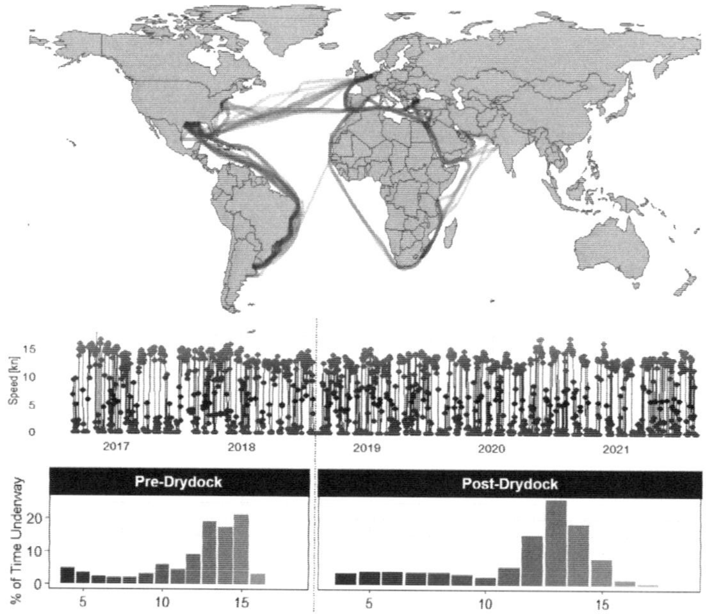

*SST refers to sea surface temperature, IWC to in-water cleaning.

Figure 5: Operational history and specification summary

The presence of large polish through areas may be related to a number of factors, such as:

- schemes not being optimised for the operational profile of the vessel,
- changes in operational profile during the inter-docking interval,
- unexpected / excessive in-water hull cleaning,
- unexpected arrival condition at the previous dry dock (additional surface preparation required) but no specification revision in dry dock to account for the changes,
- poor application control.

An analysis of the supporting information and the operational history of the vessel revealed that the specified/anticipated vessel activity never really matched the actual operational pattern of the ship – before or after dry docking. It is likely that this caused or at least contributed to the arrival condition and the level of polish through.

Issues with specifications and prevalence of defects that affect in-service performance have not been studied in detail due to lack of independent data, however, a recent study by Safinah Group based on a smaller sample of ships (180) with different technologies revealed that about 18% of the vessels included in the study had some level of polish through observed on arrival in dry dock. About 15% of the vessels arrived in dry dock with significant areas (categorised as greater than 20% of the submerged area of the hull) of 'polish through' (Figure 6) that were often covered in mature biofouling.

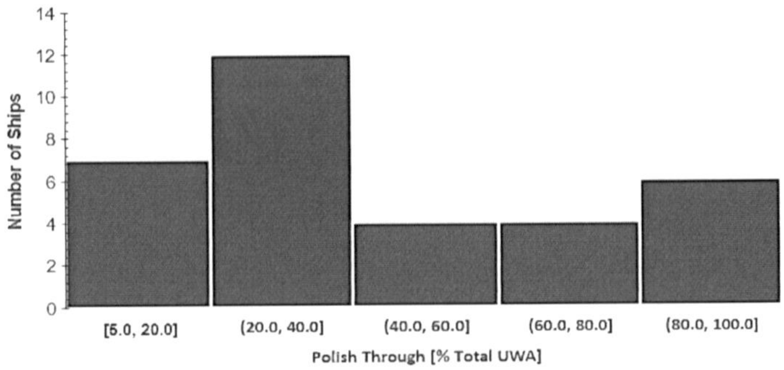

*UWA refers to underwater or submerged areas of the hull

Figure 6: Polish through presence observed on arrival in dry dock

It is important to note that premature polish through is just one defect that may directly impact long-term in-service performance that can arise from poor specifications.

6. COMMON ISSUES WITH SPECIFICATIONS AND WAYS TO ADDRESS THEM

Based on more than 512 comprehensive technical specification reviews as part of systematic coating selection and pre-planning activities and nearly as many specifications scanned for obvious errors prior to dry docking supervision activities on behalf of clients[1] since 2021, the most common issues detected are:

- Incomplete / inaccurate information provided by the client to the manufacturer or assumed by the manufacturer regarding factors needed for recommending optimal solutions and developing specifications tailored to individual ships, such as operational parameters, surface preparation, surface area size, products applied, etc;
- Errors in tender offers / specifications – products, operational profiles, surface preparation, schemes, surface areas, etc;
- Inadequate processes to accommodate potential changes in actual versus planned condition.

Apart from the issues listed above, the effects of 1) commercial pressures often leading to cost optimisation over performance considerations due to a lack of strategic alignment between stakeholders and 2) tender strategies based on specific product offerings may have a significant impact on the suitability of the recommended and chosen systems and schemes.

Many of the issues discussed above can be limited by employing a systematic approach to pre-planning of hull coating activities. Some of the aspects to be considered include:

- Technical aspect:

[1] Clients refers to entities requesting hull coating pre-planning, dry docking supervision activities or failure investigations, such as owners, operators, managers, charterers, insurers, etc.

   - A methodical analysis of the factors related to the ship's operation and history to inform specification development;
   - Ship specific technical review of tender offers ideally supplemented by independent coating in-service performance data to inform the choice of a biofouling control system/s;
   - Specification error scanning;
- Strategic and commercial aspects:
   - Align stakeholders' objectives throughout the process;
   - Ensure transparent and comprehensive record- and evidence-keeping;
   - Set up regular fleet performance reviews to create a feedback loop to inform pre-planning.

7. CONCLUSION

Optimal biofouling control is achieved through informed specification and product selection, controlled application and installation, monitoring and appropriate maintenance. Due to the number of coating technologies on the market and the lack of independent data on the in-service performance of coatings, choosing a suitable system or a combination of systems is difficult. Although a large number of decision-makers select high-technology-level options, there are additional factors, such as ensuring the specification is appropriate and tailored to a specific ship, that can limit sub-optimal performance.

This paper provides information on some of the most common issues encountered with biofouling control coating specification and selection practices and proposes a number of process modifications that can aid the industry in limiting the number of factors that may adversely affect in-service performance.

8. ACKNOWLEDGEMENTS

The authors wish to thank Mr Alan Guy and Mr Andy Hopkinson for reviewing the manuscript and Safinah Group's technical coating project supervision team for contributing to the dataset used to investigate current coating selection and performance practices.

9. REFERENCES

1. Evert A. Bouman, Elizabeth Lindstad, Agathe I. Rialland, Anders H. Strømman,State-of-the-art technologies, measures, and potential for reducing GHG emissions from shipping – A review, Transportation Research Part D: Transport and Environment, Volume 52, Part A, 2017, Pages 408-421, ISSN 1361-9209, https://doi.org/10.1016/j.trd.2017.03.022

2. IMO Circular MEPC.1/Circ.896, 2021 Guidance on Treatment of Innovative Energy Efficiency Technologies for Calculation and Verification of the Attained EEDI and EEXI, International Maritime Organization, London, United Kingdom. 14 December 2021.

3. GloFouling Partnerships Project, 'Preliminary Results: Impact of Ships' Biofouling on Greenhouse Gas Emissions', 2021.

4. IMO Resolution MEPC.207(62) 2011 Guidelines for the Control and Management of Ships' Biofouling to minimize the Transfer of Invasive Aquatic Species, International Maritime Organization, London, United Kingdom. 15 July 2011.

5. Safinah Group, 'Biofouling in Commercial Shipping: The Importance of Ship-Specific Functional Specifications', Data Insights Paper, 2021

10. AUTHORS BIOGRAPHY

Dr Ralitsa Mihaylova is the Head of Special Projects for Safinah Group, a leading independent coating consultancy based in the United Kingdom. Her background is in shipping business and operations with experience in data analysis and machine learning techniques. Ralitsa has a keen interest in sustainable solutions and the regulatory framework governing international shipping. She is actively participating in industry-led initiatives and working groups related to biofouling. At Safinah Group, Ralitsa is responsible for coordinating the company's research and development activities and delivering data-driven solutions and insights to inform strategic decisions.

Carl Barnes is General Manager Marine Consulting at Safinah Group. He joined the coating industry as a graduate chemist working in technical support and new product development. Since then, he has had extensive experience in the marine coatings industry, in technical, commercial, and marketing roles. He has a specialist knowledge of marine fouling control and regulatory aspects. Prior to joining Safinah Carl was Global Marketing Manager in Marine Fouling Control for one of the world's largest coatings companies.

Martin Porsbjerg joined the marine coatings industry as a chemical engineer / environmental chemist working in regulatory affairs and product approval for marine antifouling products. Martin has been involved in antifouling product development as well as product and technical support management for marine coatings. With 20+ years experience in the industry, he has specialist knowledge of marine antifoulings, epoxies and tank linings, standards and procedures for surface preparation and coating application in the fields of dry dock/maintenance and new building.
Over the last 6 years Martin has been heading the Safinah Marine/Technical Services Department including a large team of field-based technical coating supervisors undertaking coating condition inspections, failure investigations, pre-planning/work scope planning and execution/supervision of vessel dry dock and coating refurbishment projects on behalf of major owner/operator groups.

THE FASTWATER DEMONSTRATOR: RETROFITTING A PILOT BOAT TO METHANOL OPERATION

Y Pu, Ghent University, Belgium
P Molander and **Y Lee,** ScandiNAOS AB, Sweden
A Wiström and **A Hagander** Swedish Maritime Administration, Sweden
J Ellis SSPA, Sweden
S Verhelst Lund University, Sweden and Ghent University, Belgium

SUMMARY

Methanol as a critical piece of the puzzle for realizing net-zero shipping is a vision shared by the FASTWATER consortium, aiming to set a "FAST Track to Clean and Carbon-Neutral WATERborne Transport". This is a project funded by the European Union's Horizon 2020 framework program, which aims at demonstrating the entire value chain for using methanol as a marine fuel. This is being done by the development of retrofitted marine engines that operate on methanol and the demonstration of methanol-powered vessels that fulfil all the certification requirements. This paper focuses on the first successful demo vessel of FASTWATER: the pilot boat that is already in service in Oxelösund, Sweden, which is equipped with a compression-ignition (CI) methanol engine. The vessel's drivability and performance are proving very similar to conventional diesel engines, and its engine is certified for the International Maritime Organization (IMO) Tier III NOx standard without an after-treatment system.

1. INTRODUCTION

According to the latest greenhouse gas (GHG) study from IMO [1], waterborne transport accounts for about 3% of anthropogenic greenhouse gas emissions, and it is of no doubt that moving away from its reliance on fossil-based fuels is the way forward to mitigate this. Methanol is gaining traction as a future-proof marine fuel because it can be produced via multiple sustainable routes; it is undemanding to store either on-board or on-shore thanks to its liquid state under ambient conditions; last but not least, its production and applications can be scaled relatively easily since no scarce materials are needed and it is already traded worldwide as an important chemical feedstock.

2. METHANOL AS A FUEL FOR MARINE APPLICATIONS

2.1 METHANOL AS A FUEL FOR INTERNAL COMBUSTION ENGINES

The advantages of using methanol as a fuel for internal combustion engines have been well summarized by Verhelst et al. [2]. In short, thanks to the higher flame speed, methanol can be combusted faster than conventional fossil fuels like gasoline or diesel, which leads to lower heat loss and thus higher efficiency; thanks to the high heat of vaporization, methanol helps cool down the combustion chamber, which leads to lower pollutant NOx formation; thanks to the absence of a carbon-carbon bond and the oxygen atom bonded to its only carbon atom, methanol combustion produces no soot, which is extremely harmful to human health and also contributes significantly to global warming [3]. Methanol combustion also faces its challenges: its high autoignition temperature and corrosive nature hinder it from being a drop-in fuel for the mainstream CI marine engine. However, most of these challenges can be tackled by minor modifications to the existing marine engines. As detailed by Dierickx et al. [4], the most common method for such conversion is to ignite the methanol mixture with a small fraction of diesel injection, the so-called dual-fuel operation. Methanol can also be mixed with ignition improver and injected directly into the cylinder so that it auto-ignites at lower temperature without the help of additional diesel injection. The possibility to directly ignite neat methanol with compression at very high temperature is also under research by Svensson et al. [5].

2.2 RENEWABLE METHANOL PRODUCTION

On industrial scale, methanol is produced from the reaction of syngas over catalysts like copper and zinc oxides supported on alumina. As of today, syngas is predominantly coming from SMR (Steam Methane Reforming) with natural gas or coal gasification. Although combusting fossil-based methanol in engines yields the same efficiency and emission advantages mentioned above, it does not contribute anything in terms of GHG reduction. It was estimated that the lifecycle carbon footprint of methanol made from natural gas is as bad as conventional marine fuels like diesel or heavy fuel oil, and methanol made from coal can even be 3 times worse than conventional marine fuels [6]. Fortunately, there are multiple sustainable routes to produce methanol [7]. For example, the feedstock of SMR reactors can be changed from natural gas to biomethane from anaerobic digestion, and the feedstock of gasifiers can be changed from coal to biomass such as forestry and agricultural residue or municipal solid waste; these processes will produce bio-methanol with much lower GHG emissions. Renewable methanol can also be produced from the reaction of carbon dioxide captured from the atmosphere and the hydrogen produced from the electrolysis of water using electricity. The production of this so-called e-methanol can have an even lower lifecycle carbon footprint than bio-methanol if the electrolysers are powered by renewable electricity from wind and solar. Moreover, a recent study found that these two production pathways can be complementary [8]. Integrating electrolytic hydrogen into biorefineries enables a significant increase in the carbon conversion from the biomass feedstock to methanol which can be extra beneficial when the biomass feedstock is limited.

2.3 RENEWABLE METHANOL AS A MARINE FUEL

Stena Germanica, a large RoPax (roll-on/roll-off passenger) ferry that has been sailing between Gothenburg and Kiel since 2015, was the first commercial ship in the world to operate with methanol as its main fuel. From 2016, Waterfront Shipping commissioned 11 methanol-fuelled tankers, and 8 more are on the order book. Maersk, one of the largest shipping line and vessel operators in the world who has already committed to a net-zero emission target for 2040, has in total ordered 19 methanol-powered container ships. Thanks to these heavyweight first movers in the industry, other players are following suit. This can be seen from the latest Maritime Forecast from DNV [9] showing the number of methanol-powered ships on order to be more than triple the one in operation. However, these players do not simply jump on the bandwagon, in fact, there are already multiple studies that showcased the multitude of advantages of using methanol as marine fuel. Korberg et al. [10] has done a techno-economic assessment of advanced fuels for marine applications, where running on renewable methanol, be it made from biomass or renewable electricity, exhibits the lowest total cost of operation. Brynolf et al. [11] also demonstrated from their extensive literature study that renewable methanol has the potential to achieve lowest carbon abatement costs for ocean transport, which means it is the most economical choice to reduce GHG emissions. Lloyd's Register [12] provided a progress update of renewable ammonia, hydrogen, and methanol from different perspectives, such as technology readiness level, commercial maturity, regulatory development, and social acceptance. Renewable methanol outperforms the other two candidates due the relative simplicity for bunkering facilities, ship storage and propulsion. All these studies justified the choice made by those first movers.

3. FASTWATER

Based on the successful experiences of previous projects such as SUMMETH, GreenPilot and LeanShips, the FASTWATER consortium was formed in 2020 with the shared confidence in the potential of renewable methanol as a future-proof marine fuel. The consortium includes the entire value chain including a methanol producer, shipyards, ship owner, engine manufacturers, equipment supplier, classification society, port authority, and research institutes. As suggested by the full name of the project, "FAST Track to Clean and Carbon-Neutral WATERborne Transport through Gradual Introduction of Methanol Fuel: Developing and Demonstrating an Evolutionary Pathway for Methanol Technology and Take-up", the FASTWATER consortium aims to establish a fast transitionary path to steer waterborne transport away from fossil fuels and reduce its pollutant emissions to zero impact by using renewable methanol as a fuel. This will be achieved by the development of the relevant technologies and the demonstration of the methanol-fuelled vessels. Within the 4-year duration of the project, four vessels will be demonstrated as shown in Figure 1.

Figure 1. Four demonstrator vessels within the scope of FASTWATER (a) The pilot boat Pilot 120SE from the Swedish Maritime Administration. (b) The dual-fuel harbour tugboat from Port of Antwerp-Bruges. (c) The dual-fuel coast guard vessel from Super Toys. (d) The conceptual study of a dual-fuel river cruiser from Meyer Werft.

The one on the left is the topic of this paper, pilot boat Pilot 120SE from the Swedish Maritime Administration that runs on almost 100% methanol. It was launched in December 2021 and is in regular service since April 2022. The one on the top right corner is a harbour tugboat from Port of Antwerp-Bruges, which is currently being retrofitted to methanol/diesel dual-fuel operation. The one on the middle right is a coast guard vessel that will be built by a Greek company, Super Toys, also powered by a methanol/diesel dual-fuel engine. The one on the bottom right corner is a river cruise ship from Meyer Werft that will be demonstrated as a conceptual study for such a ship operating on methanol.

It can be clearly seen from the types of these demonstrators that FASTWATER is keen to fill the gap on the market by focusing more on retrofitting 4-stroke engines of various sizes as most of the methanol-fuelled frontrunners mentioned above are new vessels powered by newbuilt large 2-stroke engines (apart from the Stena Germanica). The pilot boat that is presented in this paper is a perfect example that renewable methanol is definitely not just a fuel for ocean-going vessels.

4. MD97 ENGINE DEVELOPMENT

4.1 BACKGROUND

As described above, dual-fuel operation is the most common technology to introduce methanol in a retrofitted diesel engine. This technology is used on the Stena Germanica and also the large two-stroke engines that currently powered the tankers from Waterfront Shipping and will power the newbuild container ships from Maersk. However, this is not the technology used in the engine presented in this paper. Instead, the engine directly ignites the mixture of methanol and ignition improver with compression, like in a traditional diesel engine. Using methanol with ignition improver as the single fuel has several advantages over dual-fuel. First, Pilot SE120 is not required to have two fuel tanks for both methanol and diesel, which can be challenging to fit into a vessel of such size. Secondly, methanol mixed with ignition improver can achieve higher methanol energy fraction than most of the methanol dual-fuel engines, which means it has higher potential to reduce GHG emissions if methanol is of renewable origin. Last but not least, alcohol fuel with ignition improver is a well-developed and mature technology. ED95, a recognized fuel standard, consisting of 95% ethanol and 5% ignition improver, was developed by Akzo Nobel and Scania during the 1980s, and Scania has produced and sold buses and trucks powered by ED95 since 1986. ED95 was developed at a time when engine control was limited to mechanical fuel injectors without the possibility to utilize pilot injection, vary pressure or injector timing. Today this is possible, which gives good control of the combustion process. To make this new methanol blend an attractive solution, the fraction of ignition enhancer should be minimized, both from a price and logistics point of view. Instead of copying the ED95 recipe, a new fuel standard should be identified that is suitable for the technology available today. The concept of a CI methanol engine was first tested in a marine application during the GreenPilot project in 2018. At the time, a 13L, 6-cylinder engine was converted and tested. In the FASTWATER project, the concept is taken to the next level by minimizing the amount of ignition improver and optimizing the emissions and power output for a commercial installation.

4.2 HARDWARE

The base of the engine is the Scania DI16. To ensure the combustion stability, the engine is mechanically modified to have a higher compression ratio in comparison to a conventional diesel engine. By replacing the pistons, the compression ratio is increased to approximately 25:1. A high-end common rail fuel system is used. The injector seat and needle are specially made for alcohol fuels. Common rail gives the possibility to inject fuel at any time, independent of the crankshaft position. The injector can inject fuel in sequences, where typically a pilot is used prior to the main injection to heat the cylinder. The purpose of a pilot injection is sometimes to reduce emissions, but it can also be used to lower engine noise. The ECU is also replaced. ScandiNAOS fine-tuned the calibration for this specific engine and methanol blend. It is prepared for propulsion, but it can also be used as a generator for auxiliary power. Table 1 details the specification of the original base engine and the converted one on Pilot 120SE. From the outside it is difficult to distinguish the MD97 engine from the original diesel engine as shown by Figure 2.

Table 1. Comparison between original DI16 and converted engine

	Original Diesel DI16	Methanol DI16
Cylinder arrangement	V-8	V-8
Displacement	16.4L	16.4L
Bore	130mm	130mm
Stroke	154mm	154mm
Compression ratio	16.7	24.7
Fuel system	Common rail diesel -type	Common rail alcohol-type
ECU	SCANIA S8	ScandiNAOS ECU & calibration

Figure 2. Methanol engine without spray-shield.

4.3 MINIMIZING IGNITION IMPROVER

Pure methanol is reluctant to auto-ignite by compression and it has poor lubricity. To modify the behaviour of methanol and make the fuel behave like a diesel fuel, it needs to be mixed with a fraction of ignition improver and lubrication additive. The ignition improver is more expensive than the methanol and thereby there is an incentive to minimize the mixing ratio; typically it accounts for about 10-20% of the total fuel price. This means that the engine needs to be 10-20% more efficient than a 100% methanol engine to motivate the higher price. With lower content of ignition improver, the methanol will be more reluctant to ignite by compression. Two critical situations were identified, cold start and step load changes, where quenching of the flame can be a potential problem. Tests onboard the boat and on dyno were prepared by mixing batches of methanol blends where the mass fraction of ignition improver varied between 1-3% in order to find the lowest content of ignition improver to achieve a good start at low ambient temperature and the step load changes up to 25% of the load without misfiring. The ignition improver used is Beraid 3555M, which is a non-ionic polyethylene glycol derivative, specially developed as an ignition improver in alcohol fuels. In the blend there is also lubrication additive, Armolube 211, that always accounts for 0.1% of mass fraction regardless of the fraction of ignition improver.

The results indicated that 3% of ignition improver is required to avoid misfiring at step loads. From tests onboard the boat, it was found that cold start was possible with 2% ignition improver, but misfire occurred in some of the step load tests. From tests on the dyno, it was found that 3% ignition improver is required to obtain the desired ignition properties. There were clear problems when using less ignition improver, which resulted in misfiring and high levels of unburnt fuel, indicated by high CO and high HC emissions. Similar to ED95, this new fuel is hence named MD97 to signify that it is methanol mixed with 3% mass fraction of ignition improver.

Within FASTWATER, there is ongoing research work on how to reduce the amount of or even completely get rid of ignition improver. The experience of the development process described above gives a very valuable input in that evaluation under steady state conditions is not enough: load variation and cold start are two critical situations that need to be addressed for the methanol engines to have stable performance.

4.4 OPTIMIZING THE EMISSION AND PERFORMANCE

The original engine installed on the pilot boat is a Caterpillar engine with a rated power output of 515kW at 2200RPM. The goal was to have similar performance to the original engine. However, physical factors will limit the engine performance to guarantee durability. These limits are cylinder pressure, exhaust gas temperature, fuel pressure and duration of injection. The goal of the dyno calibrations was to deliver more than 400kW and achieve the strictest international emission requirement, IMO Tier III. Since the engine was to be installed as a propulsion installation, the certification emissions cycle is ISO 8178 E3. NOx emission is regulated by the IMO NOx Technical Code 2008. The speed of the MD97 engine exceeds 2000RPM, which is classified as a high-speed type, so the NOx limit is 2g/kWh.

The calibration process to meet high efficiency and low emissions is iterative and the parameters of the calibration are fuel pressure, injection timing and pilot/post injections. By varying these parameters, emissions and efficiency can be optimized. The engine was first calibrated to be drivable. This was done by following a defined maximum load curve and optimized for response without any signs of high CO or HC, which could indicate poor combustion. The second step was to optimize the fuel efficiency for any load point on the engine torque/speed map. Finally, the emissions at the defined load points were optimized to meet the requirements of regulations.

The performance and emission results are summarized in Table 2. In the end, the engine itself fulfils the IMO Tier III standard without any further calibration other than drivability and fuel efficiency, which means that there are no penalties on fuel consumption to meet the strictest emission requirements. Moreover, since methanol does not form soot during combustion, and the fuel does not contain sulphur, hence there are also negligible particulate matter (PM) and zero sulphur oxide (SOx) emissions. In contrary to conventional marine diesel engines which need complicated aftertreatment system and sophisticated calibration to fight the NOx-PM trade-off, this is a significant improvement. The maximum power reached is around 415kW and the specific fuel consumption is 437-447 g/kWh, which correspond to 42% engine efficiency. This is in line with other marine diesel engines on the market. However, due to the lower rated power output from the engine, the maximum speed of Pilot SE120 is estimated to drop from 21 knots to 19 knots (which is still plenty given relevant speed limits). Figure 3 shows the power curve of the MD97 engine after the calibration and optimization process on top of the measured propeller curve of Pilot 120SE, the 4 test points defined by ISO 8178 E3 are also denoted.

Table 2. Emission result of MD97 engine.

Test point	Engine power		Engine speed		Fuel consumption	NOx	HC	CO
Unit	%	kW	%	RPM	g/kWh	g/kWh	g/kWh	g/kWh
1	100	415	100	2100	447	2.2		
2	75	311	91	1911	437	2.0		
3	50	208	80	1680	437	1.7		
4	25	104	63	1323	442	2.0		
Total						2.0	0.5	0.6

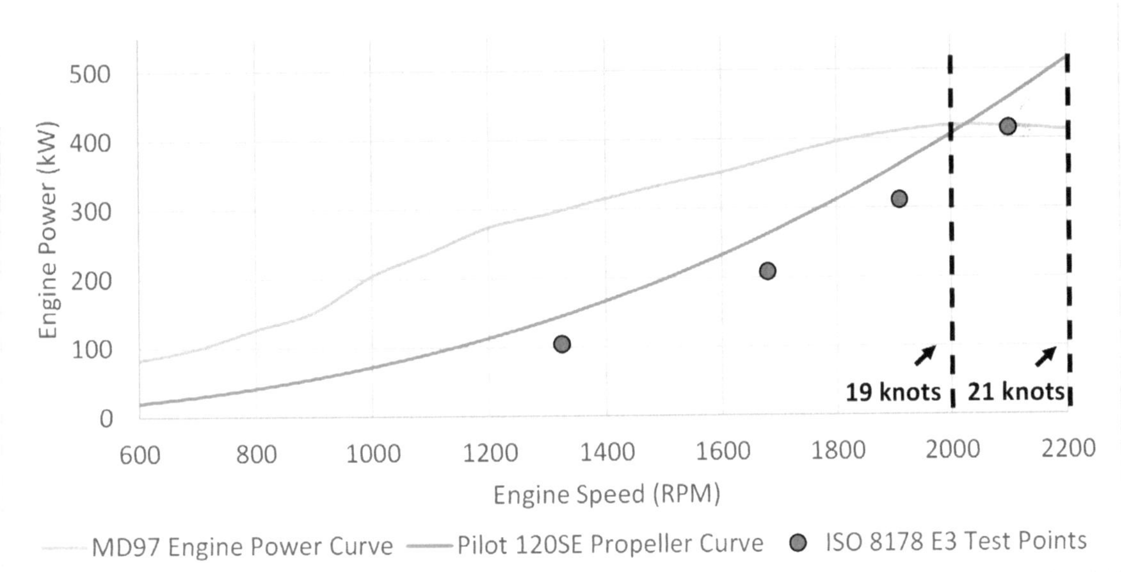

Figure 3. The power curve of the MD97 engine and the propeller curve of Pilot 120SE.

4. PILOT BOAT TECHNICAL DESIGN

4.1 BACKGROUND

LNG has emerged as a transition fuel in the last decade. Before that, the rules and regulations from authorities and classification societies were predominantly based on the characteristics of oil-based fuels since the majority of the conventional marine fuels are oil based. One example is that the flash point of a fuel must be greater than 60°C to fit into the conventional regulations. For fuels with flash point less than 60°C, special considerations need to be taken. Methanol has a flash point of approximately 11°C. To address the special characteristics of fuels with flash point lower than 60°C, IMO has issued the IGF code. The IGF code was released in 2017 but addresses so far only LNG. The work to include methanol is ongoing. In this process the IMO Maritime Safety Committee has officially approved the "interim guidelines for safety of ships using methyl/ethyl alcohol as fuel" at its 102nd session. The pilot boat is not an IMO vessel and formally, does not have to meet the IMO requirements. However, the national rules applicable for the pilot vessel will to a large extent follow the IMO rules. These rules issued by authorities and classification societies are important to ship designers. They help the designer to prepare a safe design and they represent the build-up of knowledge and experience over time. The current interim guidelines are a good start, but they still lack feedback from real applications. For instance, some sections in the guideline are very similar to the rules applicable to gas-fuelled ships, which may be overly conservative for methanol since it is a liquid. In addition, the guideline set fixed dimensions in many design criteria, some of them are hard to follow and may not be relevant for smaller vessels. Methanol, as a clean burning fuel that quickly biodegrades and thus is less harmful to marine life, and a liquid at normal temperature and pressure, is very suitable for smaller vessels that have difficulty fitting LNG tanks. By retrofitting this pilot boat, FASTWATER also aims to provide practical feedback to the authorities and improve the guidelines to ensure that the introduction of methanol technologies would not be hindered.

4.2 METHANOL FUEL SYSTEM

Overall, apart from the engine that was detailed in the previous section, methanol is also present in the bunker station, bunker lines, fuel tank, and fuel lines in the engine room. All routing of piping inside the boat is following the concept of "second barrier". This means that fuels pipes in the machinery space and in the rudder, room are double walled. Inside the fuel tank cofferdam space, the fuel pipes are single walled since the cofferdam acts as the secondary barrier. The bunker station is positioned on the aft deck and is naturally ventilated to open air. To minimize the risk for leakage the bunker connection is equipped with a drip-free coupling. Suitable materials like stainless steel are selected for components in contact with methanol. A new stainless steel methanol fuel tank replaces the original diesel tank in the rudder room. The methanol fuel tank is designed as "a tank in a tank". The inner tank is as "traditional" fuel tank, the outer tank is a second safety barrier. The inner tank has a total volume of 1.85 m^3 but considering overflow, venting and trim, the actual capacity is around 1.6 m^3. With this capacity, the pilot boat has an autonomy of approximately 150 nautical miles, at maximum speed, after conversion. The IMO interim guideline prescribes an inert gas system for blanketing the methanol tank. This might be justified for large ships with large fuel tanks of hundreds or thousands of m^3 where potentially significant volumes of methanol vapor can be formed. For smaller vessels with small fuel tanks of a couple of m^3 the situation is very different. An alternative design is applied for smaller vessels like Pilot 120SE instead of following the IMO interim guidelines. The concept is based on the "flammable triangle", by taking away one corner the tank is safe. This is a common method for land-based applications. A flame arrestor is installed in the tank ventilation, so that no sparks from outside can enter the fuel tank. All equipment used inside the tank and inside the fuel cofferdam space is EX-class. The tank is directly connected to the ship structure, so no static electricity build-up is possible.

The space between the inner and outer tank is described as a "fuel cofferdam space", which is where fuel filters, automatic fuel valves and fuel level sensors are located. The space is ventilated with extraction ventilation. Close to the ventilation outlet, a gas detector is installed. The fuel cofferdam space and the engine hood are equipped with ventilation fans with suitable capacity. The ventilation for the fuel cofferdam space has its inlet on the starboard side and its outlet on the port side of the wheelhouse. The engine hood ventilation system consists of two fans, one for operation and one for standby. To detect methanol leakage, methanol gas detectors are arranged in the engine room and fuel cofferdam space. As part of the engine safety concept, a detector is placed in the engine hood ventilation system. In case of any leaks from the engine, the methanol vapours will be extracted by the ventilation air and detected by the detector in the ventilation duct. Should a methanol leak occurs, it must be contained. To collect possible leakage of methanol, a portable tank is placed next to the engine where drainpipes direct the potential leakage to this tank. A level indicator monitors the level and gives an alarm when this portable tank needs to be replaced. Two spare portable tanks are placed on the deck that are pre-filled with water so that methanol entering them will be diluted to non-hazardous concentrations. A heat detector supplements the existing smoke detector inside the engine room to detect a methanol fire. The engine room bulkhead between rudder room and engine room is insulated with A60 insulation. To suppress a methanol fire in the engine room and bunker station, portable ARAFFF (Alcohol Resistant Aqueous Film Forming Foam) fire extinguishers are arranged. The existing Inergen (gas mixture of nitrogen, argon, and carbon dioxide) system for the engine room is upgraded to increased capacity.

4.4 BUNKERING MD97 TO PILOT 120SE

For small vessels such as a pilot boat or pleasure craft, self-fuelling stations with onsite storage can be used. These are often found in marinas and are nearly identical to stations dispensing gasoline and diesel fuel to cars. In most cases, methanol can use the same storage tanks after a proper cleaning and change of gaskets, fuelling lines, seals, and other components to accommodate methanol. National and local regulations will determine the additional safety equipment that is required, likely in conjunction with a risk assessment. The fuelling station may need to be fitted with suitable fire extinguishers such as alcohol resistant foam or dry powder, spill containment, absorbent, provision of water, emergency shower and eyewash station at the quayside, and more.

For the FASTWATER Pilot 120SE demonstrator case, a small fuel dispensing station with an integral above ground storage tank for methanol, as shown in Figure 4, was constructed by Malte Fuel & Wash. Fuel additives (ignition improver and lubricant) will be dosed during dispensing to bunker the said MD97 fuel. The methanol tank capacity of Pilot 120SE is around 1.6 m^3 so can be bunkered from the station several times between truck deliveries of methanol. In Sweden, construction and operation of fuel dispensing installations are governed by national regulations pertaining to handling and storage of flammable and explosive substances. A permit is required for the installation and a risk analysis should be submitted for the permit application. In addition, EU ATEX (Appareils destinés à être utilisés en ATmosphères EXplosibles) directives on identification and classification on hazardous zones and equipment and protective systems in identified zones must be followed.

Figure 4. MD97 fuel dispensing station for bunkering Pilot 120SE.

The MD97 dispensing station installed at the Oxelösund pilot boat station has a 7.19 m^3 inner methanol tank and a 2.5 m^3 additive tank for Beraid. The secondary containment is capable of holding 110% of the inner tank volume. The interstitial space is fitted with a leak detection system. The pump for bunkering the methanol to the pilot boat has a maximum flow rate of 80 litres per minute and a maximum pressure of 3.5 bar; the pump for transferring the methanol from the tanker truck to the dispensing station uses a flow rate of 350 litres per minute. Additional features include the grounding connection for the tanker truck, drip-free connector for the hose connection at the pilot boat and a bunker station pump safety connection to the pilot boat methanol tank "high" level indicator, with automatic shut off of the pump when the level is reached.

5. CONCLUSIONS

Since its launch in December 2021, continuous evaluation has proven that the installation of the MD97 engine onboard Pilot 120SE is successful. It has been in service in Swedish winter and summer conditions from coldest days at -15°C and more than 30°C at warmest without major issues, and the working hours are still accumulating.

The effort of the FASTWATER consortium has paid off because this demonstrator is not a one-off case; the MD97 engine developed by ScandiNAOS is now available on the market, providing a solution for other smaller vessel owners who are also interested in powering their fleet with nearly 100% renewable methanol. Introducing the MD97 engine to the market also promotes the possibility to mature the regulations regarding methanol as a marine fuel. Developing regulations that address these new technologies need to be done carefully. On the one hand, it has to ensure that the introduction of new technologies does not endanger the operation, on the other hand, inappropriate regulations where the safety measures requested are irrelevant or not well-considered would risk delaying the introduction of these new and potentially better technologies. It is thus important that more and more parties involved work jointly towards safe, practical, and economical solutions and give constructive feedback to improve the interim guidelines for using methanol as a marine fuel.

It has been emphasised multiple times in this paper that renewable methanol can be a promising future-proof fuel for marine applications. However, compared to nearly 300 million tons of fossil-based bunker fuel consumed every year worldwide, the production of renewable methanol today is unfortunately only an almost negligible number. This also prompted the "chicken or the egg" problem that newcomers are not willing to commit to methanol technologies since there is no renewable methanol available on the market. Maersk has set a good example for solving this dilemma by engaging in strategic partnerships with renewable methanol producers across the globe to secure the fuel supply for their newbuild ocean-going vessels. Maersk is, of course, a giant on the market but this strategic partnership can also be established between smaller fleet owners and renewable methanol producers on a smaller scale. According to the Methanol Institute [13], there are at the moment more than 80 renewable methanol projects globally, which are expected to produce more than 3.4 million metric tons of bio-methanol and more than 4.6 million metric tons of e-methanol annually by 2027, these increase total renewable methanol production more than tenfold compared to 2022. Balancing the demand and supply of renewable methanol will be imperative to maintain this momentum and ensure the role of renewable methanol in realising net zero shipping by 2050.

6. ACKNOWLEDGEMENTS

This paper reports part of the FASTWATER project, which stands for "FAST Track to Clean and Carbon-Neutral WATERborne Transport through Gradual Introduction of Methanol Fuel: Developing and Demonstrating an Evolutionary Pathway for Methanol Technology and Take-up". The project has received funding from the European's Horizon 2020 research and innovation program (Contract No.:860251).

7. REFERENCES

1. INTERNATIONAL MARITIME ORGANIZATION, Fourth IMO Greenhouse Gas Study, 2021

2. VERHELST, S., TURNER, J. W. G., SILEGHEM, L. & VANCOILLIE, J., 'Methanol as a fuel for internal combustion engines' *Progress in Energy and Combustion Science 70: 43-88*, 2019

3. BOND, T. C., DOHERTY, S. J., FAHEY, D. W., FORSTER, P. M., BERNTSEN, T., DEANGELO, B. J., FLANNER, M. G., GHAN, S., KÄRCHER, B., KOCH, D., KINNE, S., KONDO, Y., QUINN, P. K., SAROFIM, M. C., SCHULTZ, M. G., SCHULZ, M., VENKATARAMAN, C., ZHANG, H., ZHANG, S., BELLOUIN, N., GUTTIKUNDA, S. K., HOPKE, P. K., JACOBSON, M. Z., KAISER, J. W., KLIMONT, Z., LOHMANN, U., SCHWARZ, J. P., SHINDELL, D., STORELVMO, T., WARREN, S. G., & ZENDER, C. S., 'Bounding the role of black carbon in the climate system: A scientific assessment', *Journal of Geophysical Research: Atmospheres, 118(11): 5380-5552*, 2013

4. DIERICKX, J., BEYEN, J., BLOCK, R., HAMROUNI, M., HUYSKENS, P., MEICHELBÖCK, C., & VERHELST, S., 'Strategies for introducing methanol as an alternative fuel for shipping'. *7th Transport Research Arena TRA,* 2018

5. SVENSSON, M., TUNER, M., & VERHELST, S., 'Low Load Ignitability of Methanol in a Heavy-Duty Compression Ignition Engine', *SAE Technical Paper (No. 2022-01-1093)*, 2022

6. HAMELINCK, C., & BUNSE, M., 'Carbon footprint of methanol', *https://www.studiogearup.com/the-carbon-footprint-of-methanol-produced-from-various-feedstocks/*, January 2022

7. INTERNATIONAL RENEWABLE ENERGY AGENCY, 'Innovation Outlook: Renewable Methanol', *https://www.irena.org/publications/2021/Jan/Innovation-Outlook-Renewable-Methanol, 2021*

8. CLAUSEN, L. R., 'Maximizing biofuel production in a thermochemical biorefinery by adding electrolytic hydrogen and by integrating torrefaction with entrained flow gasification', *Energy, 85, 94-104*, 2015

9. DNV, 'Maritime Forecast to 2050', *https://www.dnv.com/maritime/publications/maritime-forecast-2022/download-the-report.html*, 2022

10. KORBERG, A. D., S. BRYNOLF, S., GRAHN, M. & SKOV, I. R., 'Techno-economic assessment of advanced fuels and propulsion systems in future fossil-free ships', *Renewable and Sustainable Energy Reviews 142*, 2021

11. BRYNOLF, S., HANSSON, J., ANDERSON, J. E., SKOV, I. R., WALLINGTON, T. J., GRAHN, M., KORBERG, A. D., MALMGREN, E. & TALJEGÅRD. M., 'Review of electrofuel feasibility—prospects for road, ocean, and air transport', *Progress in Energy 4*, 2022

12. LLOYD'S REGISTER, 'Zero Carbon Fuel Monitor', *https://www.lr.org/en/marine-shipping/maritime-decarbonisation-hub/zcfm/findings/*, July 2022

13. METHANOL INSTITUE, 'Renewable Methanol', *https://www.methanol.org/renewable/*, 2022

8. AUTHORS BIOGRAPHY

Yi-Hao Pu is currently a doctoral researcher at Ghent University. He is responsible for research and development of pure methanol engine within FASTWATER. His previous experience includes industrial diesel engines and automotive hybrid powertrains development.

Patrik Molander holds the current position of technical manager at ScandiNAOS AB. He is responsible for engine development. His previous experience includes ship design, engine testing, emission measurements and on-board installations, etc.

Yu-Sheng Lee holds the current position of Naval architect at ScandiNAOS AB. He is responsible for ship design and ship rule implementation. His previous experience includes methanol ship design, class approval and management for ship-related projects, etc.

Albert Hagander is a technical manager and department supervisor at Swedish Maritime Administration working with technology related to ship operations. He has had different roles at Swedish Maritime Administration. At the department for new building, he has been responsible for documentation in the process of purchasing components and production. Later he has been responsible for constructions of new ships on site. In the role as superintendent, he continuously answered for thirty ships owned by Swedish Maritime Administration.

Albert Wiström is superintendent at Swedish Maritime Administration, responsible for several ships. He has a background at Swedish Transport Agency, where he was involved in flag state regulations. In this context he represented Sweden in IMO meetings. Example of IMO topics he answered for was the EEDI-index (GHG-emissions) and the Polar-code. He worked for the Swedish defence material handling, being involved in the lifetime extension of the ship-series CB90H.

Joanne Ellis is a senior researcher and project manager at SSPA Sweden AB. Her work in the research department focuses on risk, safety, and environmental assessment of marine transport. She has participated in several projects on the use of methanol as a marine fuel.

Sebastian Verhelst holds the current position of Full Professor at Ghent University (Belgium), and Associate Professor at Lund University (Sweden). He is responsible for research into the application of renewable fuels in combustion engines, and teaching in reciprocating machinery, combustion, and fuels. His experience includes work on hydrogen, methanol, various bio- and drop-in fuels, for realizing sustainable transportation.

THE ROYAL INSTITUTION
OF NAVAL ARCHITECTS

CONFERENCE PROCEEDINGS

THE ROYAL INSTITUTION OF NAVAL ARCHITECTS

8-9 Northumberland Street, London WC2N 5DA

Tel: +44 (0)207 235 4622 Fax: +44 (0)207 259 5912

events@rina.org.uk

www.rina.org.uk

The Royal Institution of Naval Architects
8-9 Northumberland Street
London, WC2N 5DA
United Kingdom

ISBN 978-1-7138-8302-9

9 781713 883029

CURRAN ASSOCIATES INC.
proceedings
.com

European Technical Coatings Congress (ETCC 2018)

Towards a Bright Future, New Coating Technologies

Amsterdam, The Netherlands
26 - 29 June 2018

ETCC
EUROPEAN TECHNICAL COATINGS CONGRESS